Radar Systems

Macmillan New Electronics Series
Series Editor: Paul A. Lynn

Paul A. Lynn, *Radar Systems*
A. F. Murray and H. M. Reekie, *Integrated Circuit Design*

Radar Systems

Paul A. Lynn

BSc (Eng), DIC, PhD, MIEE, C Eng
formerly Reader in Electronic Engineering, University of Bristol

Macmillan New Electronics
Introductions to Advanced Topics

MACMILLAN
EDUCATION

First published 1987

Published by
MACMILLAN EDUCATION LTD
Houndmills, Basingstoke, Hampshire RG21 2XS
and London
Companies and representatives
throughout the world

Printed in Hong Kong

Typeset by
TecSet Ltd, Wallington, Surrey

British Library Cataloguing in Publication Data
Lynn, Paul A.
    Radar systems.—(Macmillan new electronics)
    1. Radar
    I. Title   2200   9850   9351
    621.3848'5   TK6575

    ISBN 0-333-42543-X
    ISBN 0-333-42544-8 Pbk

Contents

Series Editor's Foreword

The rapid development of electronics and its engineering applications ensures that new topics are always competing for a place in university and polytechnic courses. But it is often difficult for lecturers to find suitable books for recommendation to students, particularly when a topic is covered by a short lecture module, or as an 'option'.

Macmillan New Electronics offers introductions to advanced topics. The level is generally that of second and subsequent years of undergraduate courses in electronic and electrical engineering, computer science and physics. Some of the authors will paint with a broad brush; others will concentrate on a narrower topic, and cover it in greater detail. But in all cases the titles in the Series will provide a sound basis for further reading of the specialist literature, and an up-to-date appreciation of practical applications and likely trends.

The level, scope and approach of the Series should also appeal to practising engineers and scientists encountering an area of electronics for the first time, or needing a rapid and authoritative update.

Preface

The basic principles of radar do not change, but the design and technology of practical radar systems have developed rapidly in recent years. Advances in digital electronics and computing are having a major impact, especially in radar signal processing and display. I hope that this book will prove a useful introduction to such developments, as well as to the underlying principles of radar detection.

A short book on a huge subject must be selective. I have decided to concentrate on pulse radars used for Air Traffic Control, including secondary radar. In part this reflects personal experience. But I also believe that modern ATC radar, with all its recent advances in signal processing, provides an excellent framework for introducing many of the problems and solutions of radar engineering. Civil equipments of this type are also relatively accessible, for visits or for technical information.

Readers requiring practice in radar calculations will find a set of problems (with solutions) towards the end of the book. They relate to the quantitative material developed in chapters 2, 3 and 5.

I gratefully acknowledge information obtained from many sources. Most of the references listed at the end of the book have influenced me considerably. I have also received technical material and photographs from various radar manufacturers, and should like to thank:

> Cossor Electronics Ltd (Harlow, England)
> English Electric Valve Company Ltd (Chelmsford, England)
> Marconi Radar Systems Ltd (Chelmsford, England)
> Plessey Radar Ltd (Chessington, England)
> Racal Avionics Ltd and Racal Marine Radar Ltd (New Malden, England)
> AEG (Ulm, West Germany)
> Hollandse Signaalapparaten (Hengelo, The Netherlands)
> Thomson/CSF (Meudon-la-forêt, France)
> Westinghouse Electric Corporation (Baltimore, USA)

Several national ATC authorities have been very helpful. I have used information supplied by the British Civil Aviation Authority and National Air Traffic Service in section 7.2 of the book; by the French ATC authority (Direction de la Navigation Aérienne) in section 7.3; by the West German ATC authority (Bundesanstalt für Flugsicherung) in section 7.4; and by the USA Federal Aviation Administration in section 7.5. It is also a personal pleasure to recall hospitality received during visits to radar sites near Gatwick, England; Evreux, France; and the ATC centre and radar near Hanover, Germany.

It is only fair to add that most of the information obtained from the above sources has been heavily condensed. I regret any errors of fact or balance which may have crept in. I am also very aware of topics which have had to be omitted altogether for lack of space, and trust that such omissions will not seriously detract from the book's value.

I should finally like to thank Jeana Price for all her careful work on the typescript.

4 Kensington Place   Paul A. Lynn
Clifton
Bristol BS8 3AH

List of Symbols

α	Weibull parameter
Δ	antenna difference pattern
$\delta[n]$	discrete unit impulse
ϵ	permittivity
η	clutter volume density
θ	phase or azimuth angle
θ_B	horizontal beamwidth
λ	wavelength
μ	permeability
ρ	reflection coefficient
ρ_A	antenna efficiency
Σ	antenna sum pattern
σ	target cross-section
σ_0	clutter area density
σ_{av}	average cross-section
σ_c	clutter cross-section
τ	pulse length
τ'	auxiliary time variable
ϕ	elevation, grazing, or phase angle
ϕ_B	vertical beamwidth
ψ_0	variance
$\psi_0^{\frac{1}{2}}$	standard deviation
ψ_d	phase difference
Ω	discrete frequency variable
ω	continuous frequency variable
A	antenna physical area
a	antenna aperture
A_c	clutter area
A_e	antenna effective area
$A[n]$	antenna aperture distribution (discrete)
$A(z)$	antenna aperture distribution (continuous)
B	bandwidth
C	clutter power
c	velocity of radio waves

D	Earth's effective diameter
d	distance or horizontal range
$E_i(n)$	integration efficiency
$E(\phi)$	electric field distribution
F	frequency spectrum
f	frequency in hertz
f_0	transmitter frequency
f_c	intermediate frequency
f_d	doppler frequency
f_1	stalo frequency
F_n	noise figure
f_p	pulse repetition frequency
f_r	beat frequency
G	antenna gain
G'	receiver or amplifier gain
h	height
$h[n]$	impulse response (discrete)
$h(t)$	impulse response (continuous)
$H(\Omega)$	frequency response of digital system
I	in-phase MTI channel or MTI improvement factor
I_0	Bessel function of zero order
k	Boltzmann's constant or integer variable
l	length
L_s	loss factor
N	integer number or variable
n	hits-per-target or integer variable
p	probability density
P_{av}	average transmitter power
P_d	detection probability
P_{fa}	false-alarm probability
P_r	received power
P_t	transmitter power
Q	quadrature MTI channel
R	range
r	rainfall rate
R'	path length
R_{max}	maximum range
S	signal power
S_{min}	minimum detectable signal
(S/C)	signal-to-clutter power ratio
$(S/N)_1$	input $S:N$ ratio
T	interpulse interval, sampling period, or delay time
t	time
T_0	absolute temperature or time interval/delay

t_0	time delay
v	velocity
v_1	input voltage
v_2	output voltage
V_c	clutter volume
v_r	radial velocity
$v(t)$	voltage signal
$v_e(t)$	envelope voltage signal
V_T	threshold voltage
W	power spectral density
X_k	discrete spectral coefficient
$x[n]$	input signal (discrete)
$x(t)$	input signal (continuous)
$y[n]$	output signal (discrete)
$y(t)$	output signal (continuous)

List of Abbreviations

ACF	autocorrelation function
ADC	analog-to-digital converter
ADT	automatic detection and tracking
ARSR	air-route surveillance radar
ASDE	airfield surface detection equipment
ASMI	airfield surface movement indicator
ASR	airfield surveillance radar
ATC	air traffic control
BITE	built-in test equipment
CFAR	constant false-alarm rate
COHO	coherent oscillator
CPI	coherent processing interval
CRT	cathode ray tube
CW	continuous wave
DAC	digital-to-analog converter
DFT	discrete Fourier Transform
DPSK	differential phase-shift keying
DVST	direct-view storage tube
ERP	effective radiated power
FET	field-effect transistor
FFT	fast Fourier Transform
FIR	flight information region
FM-CW	frequency-modulated continuous wave
FTC	fast time constant
GCA	ground-controlled approach
IAGC	instantaneous automatic gain control
ICAO	International Civil Aviation Organisation
IF	intermediate frequency
IFF	interrogation, friend or foe
LO	local oscillator
LTI	linear time-invariant
LVA	large vertical aperture
MTBF	mean time between failures
MTD	moving target detector
MTI	moving target indicator

NM	nautical mile
OBA	off-boresight azimuth
OTH	over the horizon
PAR	precision approach radar
PCR	pulse compression ratio
PDF	probability density function
PPI	plan-position indicator
PPS	pulses per second
PRBS	pseudo-random binary sequence
PRF	pulse repetition frequency
PSD	phase-sensitive detector
RADAR	radio detection and ranging
RHI	range–height indicator
RPM	revolutions per minute
SAW	surface acoustic wave
SCV	subclutter visibility
SD	standard deviation
SSR	secondary surveillance radar
STC	sensitivity time control
TR	transmit–receive
TWT	travelling wave tube
ZVF	zero-velocity filter

1 Introduction

1.1 Historical notes

The reflection of radio waves by conducting objects was first noticed more than a century ago. As far back as 1903, the effect was used in Germany to demonstrate detection of ships at sea. Marconi championed the same idea in Britain in 1922. However, there was little official interest and several years passed before systematic experiments in radio detection began. Early work used continuous-wave (CW) transmissions, and relied upon interference between a transmitted wave and the doppler-shifted signal received from a moving target. The detection of aircraft was first accomplished in the USA in 1930.

CW transmissions can detect the *presence* of an object. If the radio wave is formed into a narrow beam, they can also indicate *direction*. But they cannot give *range*. This serious limitation may be overcome by modulating the transmitter output in some way – for example, by sending out a train of short pulses. The time taken for echoes to return to the receiver is then a direct measure of target range. Strictly, only systems of this type should be called 'radar', which is a shortened form of *Radio Detection and Ranging*.

Practical development of pulse radar began in the 1930s – principally in the USA and Britain, but also in several other countries. Although much of the initial work was American, the British effort intensified after 1935 because of the threat of invasion. In that year Robert Watson Watt was asked by the British Government to report on the use of radio as a death ray. He concluded that this was impracticable, but recommended a major research effort into the radio detection of aircraft. By the end of 1935, ranges of over 50 miles had been achieved on bomber aircraft. This work culminated in the CH (Chain Home) radar stations which greatly contributed to the defence of Britain against air attack in 1940.

Before the Second World War, radio frequencies between about 5 and 200 MHz were used in radar. The transmitters and receivers were essentially modifications of equipment used for radio broadcasting and reception. However, it was realised that considerably higher, microwave, frequencies are desirable. The main reason is that accurate location of a target requires a narrow radio beam. This is only possible if the antenna aperture is much greater than the radio wavelength (λ). Even at 200 MHz ($\lambda = 1.5$ m) a very large antenna is needed.

1

Large structures are expensive, difficult to steer, and suffer from heavy wind loading. They are unsuitable for ships, and cannot be installed in aircraft.

By 1940 the British researchers Randell and Boot had developed their cavity magnetron. This new type of transmitter tube could deliver up to 1 kW output power at 10 cm wavelength, and made microwave radar feasible. The device was demonstrated in the USA in September 1940. From that moment onwards, the two countries collaborated closely on pulse radar research.

Both the theory and the technology of radar were developed with great urgency during the Second World War. Since 1945 the pace has been slower, but steady. Radar has found widespread civilian application, including air traffic control, meteorology, radar speed traps, and the measurement of insect swarms and crop growth. The basic principles of radar do not change, but there have been continuous improvements in hardware and system design. There has also been a major increase in the use of computers and digital techniques for radar signal processing and display. The trend seems certain to continue, and we shall have more to say about it later in this book.

1.2 Types of radar system

1.2.1 Transmission waveforms

One of the most important technical characteristics of a radar system is the type of transmission waveform it uses. There are four more-or-less distinct categories.

In a *continuous-wave (CW) radar*, the transmitter output is a continuous, unmodulated, radio-frequency oscillation. A major problem is to provide adequate screening between the transmitter and the receiver. Indeed, transmitter 'breakthrough' generally prevents the receiver from detecting echoes from stationary targets – even if separate antennas are used for transmission and reception. The receiver is instead designed to detect the doppler shift in echoes reflected from moving targets. This is done by mixing a small portion of the transmitter output (or a signal derived from it) with the received signal. The doppler, or difference, frequency is then extracted. The doppler frequency is given by

$$f_d = \frac{2v_r}{\lambda} = 2v_r \frac{f_0}{c} \tag{1.1}$$

Here v_r is the target velocity relative to the radar, referred to as the *radial velocity*; f_0 is the transmitter frequency; and c is the velocity of radio waves $(3 \times 10^8 \text{ m s}^{-1})$. Velocities of aircraft and ships are usually expressed in nautical miles per hour, or knots, where

$$1 \text{ knot} = 0.515 \text{ m s}^{-1} = 1.85 \text{ km/hour} = 1.15 \text{ statute miles/hour} \tag{1.2}$$

For example, if f_0 = 3 GHz (λ = 10 cm) and it is required to detect aircraft with radial velocities up to 500 knots (258 m s^{-1}), equation (1.1) indicates doppler frequencies in the range 0–5 kHz. Ships with radial velocities up to 30 knots would give doppler frequencies up to a few hundred hertz.

It is possible to extract the doppler frequency directly by mixing transmitted and received signals. This is known as *homodyne detection.* However, most practical radars use the *heterodyne* principle, in which incoming signals are first converted to an intermediate frequency (IF). This offers a number of advantages, including a reduction in receiver noise. Following IF amplification the echoes are detected and the doppler signals extracted. Since these signals often fall in the audio-frequency range, a pair of headphones may be used as a simple and effective receiver output device. Of course, accurate measurement of target velocity requires some form of frequency counter or display.

CW radar has found many practical applications. They include rate-of-climb meters for vertical take-off aircraft, radar speed traps, and speedometers. Broadly speaking, CW radar is best suited to inexpensive short-range systems requiring limited transmitter power. In high-power, long-range systems it becomes increasingly difficult to protect and shield the receiver from the transmitter.

We next consider *frequency-modulated continuous-wave (FM–CW) radar.* We have previously noted that a CW radar cannot indicate target range. One way round the problem is to modulate the transmitter output frequency. A triangular or sinusoidal modulating waveform is commonly used. By measuring the difference frequency, or *beat frequency*, between the instantaneous transmitter frequency and the frequency of a received echo, it is possible to infer the range of a target.

The simplest case to consider is when the transmitter carrier frequency is changed linearly with time over a certain interval, as in triangular modulation. Suppose its rate of change over a half-cycle of the modulation is df_0/dt. Now an echo from a target at range R (the go-and-return path being $2R$) experiences a time delay $T_0 = 2R/c$. If the target is stationary and there is no doppler shift, the beat frequency is

$$f_r = T_0 \left(\frac{df_0}{dt} \right) = \frac{2R}{c} \left(\frac{df_0}{dt} \right) \tag{1.3}$$

This is proportional to target range. Of course, the situation is more complicated with a moving target, since there is an additional doppler shift imposed on f_r. If the transmitter frequency is instantaneously *increasing*, the net beat frequency due to an approaching target is $(f_r - f_d)$; for a receding target it is $(f_r + f_d)$. These effects are reversed if the transmitter frequency is instantaneously decreasing. However, by averaging the net beat frequency over a complete modulation cycle, the beat frequency due to target range alone may be found. Thus

$$f_r = \frac{1}{2} \{ (f_r - f_d) + (f_r + f_d) \} \tag{1.4}$$

It may be shown that averaging also produces the correct value for f_r when sinusoidal frequency modulation is used.

Generally speaking, there is ambiguity over the interpretation of net beat frequency in an FM–CW radar. For example, a given frequency may be produced either by a fast target at short range, or by a slow target at long range. Fortunately, many practical systems are designed to detect relatively slow-moving targets at fairly long range. In this case the range component f_r is much larger than the doppler component f_d. A good example is the type of FM–CW radio altimeter commonly used in aircraft.

A different solution to the problem of range is provided by *pulse radar*. In this type of system, the transmitter sends out a train of short radio-frequency pulses. Target range is found by measuring the time for echoes to return to the receiver. Pulse radar has found extensive practical application, and we shall concentrate on it in this book. One of its major advantages is that since the transmitter is turned off most of the time, the receiver can 'listen for' returning echoes without any interference from the transmitter. The same antenna is normally used for transmission and reception.

Figure 1.1 shows a simplified block diagram of a classic pulse radar. The pulse modulator on the left-hand side generates a rectangular pulse train which turns the transmitter on and off. The transmitter output is fed to the antenna via a special switch, known as a *duplexer*. This protects the receiver input stages from an excess of transmitter breakthrough. As soon as a transmitter pulse is completed, the duplexer connects the antenna to the receiver. Returning echoes are first processed by a high-quality, low-noise amplifier. They are then converted to IF by mixing with the output of a local oscillator (LO), using the heterodyne principle. Further amplification, plus bandlimitation to reduce system noise, is performed at IF. The received signal is then detected to produce a video waveform. After final amplification and processing, it is used to drive a display.

A common form of display, used with many air-traffic-control and marine radars, is the *plan position indicator (PPI)*. This gives a 'bird's eye view' of

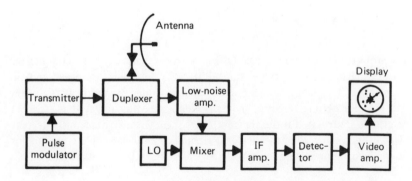

Figure 1.1   Block diagram of a basic pulse radar

the area covered by the radar. The basic type of analog PPI display incorporates a rotating trace controlled by the antenna heading. The trace, starting from the centre of the display each time the transmitter fires, travels to the edge at constant speed. The time delay of an echo, and hence target range, is therefore directly shown as radius on the display. As each returning echo is received, it causes a 'blip' on the screen.

The rest of this book will concentrate on pulse radar, so we need not go into much detail at this stage. However, it is helpful to mention just two important operating parameters of such a system: the *pulse repetition frequency (PRF)*; and the *pulse length*.

The PRF is usually made as high as possible, consistent with the need to avoid *second-time-around echoes*. In other words, sufficient time is allowed between successive transmitter pulses for echoes to return from the most distant targets. Otherwise the echoes would arrive after the transmission of the next pulse, leading to uncertainty about target range. This is essentially another aspect of the radar ambiguity problem, already mentioned in relation to FM–CW radar. Suppose we expect to detect aircraft out to some maximum range R_{max}. The echo time will be $2R_{max}/c$, so successive transmitter pulses must be spaced at least this much apart. We see that second-time-around echoes will be avoided only if the PRF is less than

$$f_p = c/2R_{max} \text{ pulses per second (pps)} \tag{1.5}$$

For example, if R_{max} = 200 nautical miles (nm), $f_p <$ 405 pps. But if the maximum range we expect is 10 nm, f_p may be increased to 8100 pps.

The choice of pulse length represents a compromise between two major factors. On the one hand, a very short pulse gives good range resolution, and allows accurate measurement of short ranges. But if it is too short, detection becomes very difficult. Practical systems generally use pulse lengths between about 0.1 and 5 μs. We should also note that modern radar systems quite often employ *pulse compression*. A much longer, coded pulse is transmitted, and the echoes are 'sharpened' in the receiver using special signal processing techniques.

The doppler-shift effect is also very important in many pulse radars. Unlike CW systems, doppler information is not continuously available, being confined within each narrow received pulse. This makes it harder to extract. Nevertheless, some highly effective methods including *moving target indication (MTI)* have been developed for discrimination in favour of aircraft and other moving targets.

Mention of moving target indication leads us directly to a fourth type of system, the so-called *pulse doppler radar*. This is not strictly a separate category, but is closely related to MTI pulse radar. As already noted, MTI pulse radar overcomes range ambiguity by using a PRF low enough to avoid second-time-around echoes. However, this makes it subject to velocity ambiguity and to the occurrence of *blind speeds*. These are particular values of radial velocity at which targets tend to become 'invisible'. A pulse doppler system achieves the opposite compromise. It avoids velocity ambiguity by using a considerably higher PRF, but is subject to range ambiguity. In practice MTI pulse radar is more common, and we shall concentrate on it in this book.

1.2.2 Operating frequencies

The transmission frequency of a radar has an important influence on its performance in adverse weather. It also affects such decisions as the size and type of antenna to be used, and the choice of transmitter and receiver hardware. In practice, wavelengths above about 50 cm produce negligible weather echoes. But wavelengths below about 3 cm may produce very severe ones. This is a consequence of the phenomenon known as *Rayleigh scattering*, whereby the visibility of a small water droplet or particle to a radar is inversely proportional to the fourth power of the wavelength. Apart from meteorological applications, a radar operator does not usually wish to see heavy echoes from clouds, rain, or snow on his display. For this reason, and also because short radar wavelengths suffer increasingly from atmospheric attenuation, most medium-range and long-range systems operate at 10 cm or above.

Although most practical radars operate at wavelengths between about 1 cm and 50 cm, there is considerable interest in systems operating outside this range. For example, millimetric radars have been developed for their very high resolution, which can be achieved using physically small antennas. At the other

Figure 1.2   The antenna of a long-range radar used by the British Civil Aviation
Authority for aircraft surveillance (photo: Paul A. Lynn)

extreme, the HF region of the radio spectrum has the special property of allowing propagation beyond the Earth's curvature by means of refraction from the iono-sphere. The resulting *sky waves* are exploited in so-called *over-the-horizon (OTH)* radars, which can give ranges of many hundreds, or even thousands, of kilometres. Shorter-range working may be obtained using *ground waves* propa-gated around the curvature of the earth by diffraction. Ranges are typically up to a few hundred kilometres. The wavelengths used in HF radar are normally between about 10 m and 100 m. Of course, at such wavelengths it is very diffi-cult to produce a narrow beam using a steerable antenna.

In the early wartime days of microwave radar, letters of the alphabet were used to denote various frequency bands. Although the original aim of preserving secrecy is no longer relevant, the same letters (L, S, C, X, K) are still widely used by radar engineers. They are indicated in the upper part of figure 1.3. A more logical series of letters, A to M, has now been introduced, and these are shown in the lower part of the figure. It must be said that there is some resist-ance to the change, many people still preferring the historical series. This can lead to a certain amount of confusion, particularly over bands C and L. The other point to make is that particular radar applications, such as civil Air Traffic Control, are generally restricted by national or international agreement to limited frequency ranges within each band.

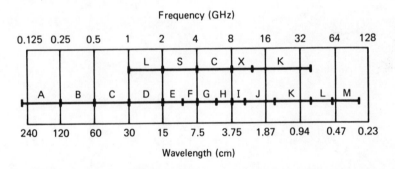

Figure 1.3 Radar frequency bands and wavelengths

1.2.3 Applications

Radar systems may also be classified according to application. We have already mentioned a number of these. It is now time to make a proper list, and intro-duce some operational terminology.

Air-traffic-control (ATC) radar

These systems are used by civilian and military aviation authorities to monitor air traffic en-route, to maintain adequate separation between aircraft, and

(a)

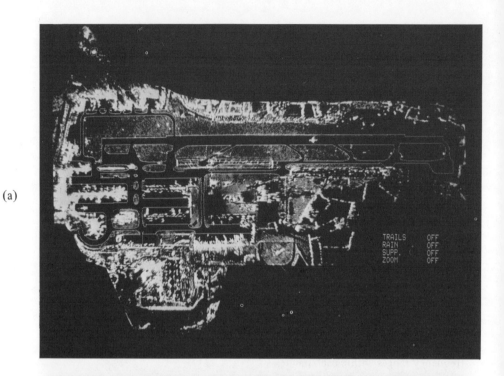

TRAILS   OFF
RAIN   OFF
SUPP.   OFF
ZOOM   OFF

(b)

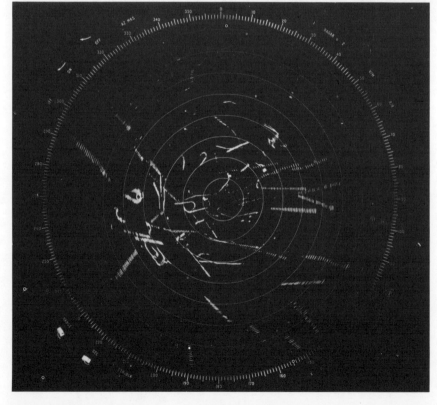

to guide them safely into and out of airports. Most ATC radars are ground-based. Within this broad category we may identify several types.

En-route surveillance radar. Such radars typically provide coverage on aircraft at ranges up to 200 nm. Since they are designed to monitor air traffic over a wide area, they are not necessarily sited close to airports. Their signals may well be sent by radio or cable to a distant ATC centre, and combined with information from other radars.

Terminal area radar. Typically providing coverage to about 70 nm, such radars are often based at or near major airports. Their main role is the guidance and separation of incoming and outgoing aircraft.

Approach control radar. This type of system requires a somewhat shorter range than a terminal area radar, and is often installed at smaller airports.

Precision approach radar (PAR). This type of radar is designed to guide aircraft to a runway threshold, especially in bad weather. Such an operation is sometimes referred to as *ground-controlled approach* (GCA). With a typical range of 15 or 20 nm, a PAR often points in the direction of a particular runway.

Airfield surface movement indicator (ASMI). Short-range, high-resolution radars are sometimes installed at airports to help monitor aircraft and vehicles on the ground, and to check that runways are clear for take-off and landing. Such systems are also called airfield surface-detection equipments (ASDEs).

Airborne radar

Radar may be installed in aircraft for several reasons, including weather avoidance, terrain avoidance in low-level flight, and ground mapping. Many of the applications are military. The need for a small, lightweight antenna favours short radar wavelengths. As already mentioned, the widely used FM–CW altimeter is also a form of airborne radar equipment.

Marine radar

Radar is widely used by ships and pleasure vessels for avoiding collision at sea, for detecting navigation markers in poor weather, and for mapping nearby coast-

---

◀ Figure 1.4 (a) PPI display of an ASMI radar at London (Gatwick) Airport. Operating at about 9.4 GHz, the system uses a narrow beam and very short pulse to achieve high resolution. Note the aircraft on the main runway (photo courtesy of Racal Avionics Ltd).
(b) Long-exposure photograph of the PPI display of a terminal-area surveillance radar, showing more than 50 aircraft tracks. The range rings are 10 nm apart. Sophisticated signal processing is needed to achieve a display of this clarity (photo courtesy of Marconi Radar Systems Ltd)

lines. Shore-based radars of similar range and performance are used by port and navigation authorities for controlling shipping in harbours and waterways. Naval vessels use a variety of radars for detecting aircraft and missiles.

Tracking radar

Radar may be used to track a moving target and to predict its future position. Most applications are military. Such a system must first search for, or *acquire*, its target before it can track it. Acquisition and tracking may be performed by separate subsystems. During tracking, the antenna beam is positioned in space by a servo driven by an error signal derived in the receiver.

Satellite and space radar

Large ground-based radars are used for satellite tracking. Space vehicles may incorporate radar for rendezvous and docking.

Remote sensing radar

Radar can give much information about the physical environment. For example, it is used in meteorology, and for measurements of sea state, rainfall, crop growth, insect swarms and environmental pollution.

Other radar applications

Further applications include speedometers, speed traps, proximity detectors, devices for collision avoidance and intruder alarms.

We therefore see that radar has found a host of practical uses. In a short book it is obviously necessary to be selective. We concentrate here on pulse radars, particularly those used for surveillance in air traffic control. Such systems present the radar designer with some of his biggest challenges, and give a good overall introduction to the radar problem.

2 The Radar Equation

2.1 Basis of the equation

In this chapter we begin a more detailed account of radar. The first task is to develop a simple form of the *Radar Equation*. This clarifies important relationships between such factors as transmitter power, receiver sensitivity, target size and range.

The reader should be clear that our initial calculations assume a very basic type of radar system. Most practical systems are much more complicated. For example, radar antennas often generate more than one beam, and their transmitters may work at more than one frequency. Pulse compression may be used in the receiver. Computers and digital techniques are increasingly used to detect and to process target echoes, and to generate a radar picture on a digital display. However, by concentrating initially on the traditional system in which returning echoes cause 'blips' on an analog display, we may obtain some valuable insights into the whole radar problem. We will, of course, consider modern developments later in the book.

Some quite simple calculations can give useful information about overall system performance. Figure 2.1 represents a transmitter–receiver, and a target at range R. We start by assuming that the transmitter power P_t is radiated *isotropically* – that is, equally in all directions. At range R the power is spread evenly over an imaginary spherical surface of area $4\pi R^2$. The power density (watts per m^2) at the target is therefore

$$\frac{P_t}{4\pi R^2} \tag{2.1}$$

A directive antenna is, of course, used in practice. We define its gain G as the increase in power density produced at the target, compared with an isotropic radiator. Thus, the actual power density at the target using a directive antenna is

$$\frac{P_t G}{4\pi R^2} \tag{2.2}$$

We next assume that the target has area σ, and reflects all incident power isotropically. The power density of the echo at the receiver is therefore

$$\frac{P_t G}{4\pi R^2} \times \frac{\sigma}{4\pi R^2} = \frac{P_t G\sigma}{16\pi^2 R^4} \tag{2.3}$$

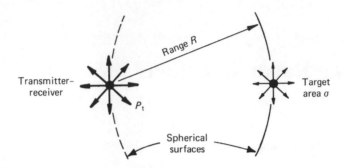

Figure 2.1 The basic geometry of radar

The reader may object that a real target is most unlikely to behave in this way. It will not re-radiate all the incident power, nor will it do so isotropically. However, this is taken into account by the definition of target area. σ is not an actual physical cross-section. It is the *effective* area 'seen' by the radar, assuming isotropic re-radiation.

If the effective collecting area of the receiving antenna is A_e, the received power is

$$P_r = \frac{P_t G \sigma A_e}{16\pi^2 R^4} \tag{2.4}$$

In practice, noise in the receiver will prevent an echo from being reliably detected if P_r falls below some minimum value. This is called the *minimum detectable signal, S_{min}*. Clearly, a target will just be detectable at a range R_{max} if the echo power it produces equals S_{min}. Substitution in equation (2.4) gives

$$S_{min} = \frac{P_t G \sigma A_e}{16\pi^2 R_{max}^4}, \quad \text{or} \quad R_{max} = \left(\frac{P_t G \sigma A_e}{16\pi^2 S_{min}}\right)^{\frac{1}{4}} \tag{2.5}$$

Before proceeding, we should emphasise that these equations do not predict the performance of practical radars with great accuracy. They are generally too optimistic. This is because radar detection is an inherently statistical problem. The certainty implied by equations (2.4) and (2.5) is not realistic. We shall have more to say about this in the following sections.

In spite of their limitations, the above results are useful for obtaining orders of magnitude. We may illustrate by an example. Suppose the receiver of a pulse radar can detect a signal power level of 10^{-13} W. We wish to know roughly what transmitter power will be needed to detect a medium-size aircraft ($\sigma = 10$ m^2) at 200 nm range. Let us assume a large antenna with an effective receiving area of 20 m^2, and a typical power gain of 1000 (30 dB). Now 200 nm equals 370 km, or 3.7×10^5 m. Rearranging equation (2.5), we obtain

$$P_t = \frac{16\pi^2 S_{min} R_{max}^4}{G \sigma A_e} = \frac{16\pi^2 \times 10^{-13} (3.7 \times 10^5)^4}{10^3 \times 10 \times 20}$$

and therefore

$$P_t = 1.6 \times 10^6 \text{ W} \quad \text{or } 1.6 \text{ MW} \tag{2.6}$$

The result correctly suggests that megawatt pulse powers may be needed in long-range surveillance applications. It is also interesting to note the enormous ratio of about 10^{19} between transmitted and received powers in this example.

Equation (2.4) underlines the dramatic dependence of received power on target range. Other factors remaining constant, the strength of an echo is inversely proportional to the fourth power of range. This is because the inverse square law of radiation operates in both directions — from transmitter to target, and from target back to the receiver. It explains why the displays of long-range systems can easily become cluttered by echoes from birds or insects at ranges of 50 km or more.

To help emphasise these ideas, the reader may like to estimate the transmitter power needed for a precision approach radar (PAR) designed to detect a light aircraft ($\sigma = 1 \text{ m}^2$) at ranges up to 15 nm. Assume a highly directive antenna of gain 36 dB, with an effective receiving area of 1 m^2. The minimum detectable signal is, once again, 10^{-13} W. The answer turns out to be about 2.5 kW — not much more than a thousandth of the power required by the long-range system. Of course, we should again emphasise that such a figure is only approximate.

A radar receiver can easily become overloaded by strong echoes from nearby terrain, objects and targets. It is therefore common practice to desensitise the receiver immediately after the transmitter fires. A convenient way of doing this is to incorporate attenuation on the input side of the receiver. During the early part of each interpulse period, the effective gain of the receiver is low. It is then allowed to recover in a controlled manner — often following an approximately fourth-power characteristic with time. By the time echoes are received from distant targets, the effective gain has reached its maximum value. This technique is known as *swept gain* or *sensitivity time control (STC)*.

This is a good moment to distinguish between *primary* and *secondary* radar. Our calculations in this section (and in most of the rest of the book) are concerned with primary radar, in which the target acts as a passive reflector. As we have seen, the inverse fourth-power relationship between echo strength and range presents a major problem for long-range detection. It is also very difficult to determine the height of an aircraft sufficiently accurately using primary radar. Secondary radar overcomes both problems by requesting an *active response* from the aircraft. It 'interrogates' the aircraft, which normally responds with both height information and an identification. The power requirements of a secondary radar transmitter are quite modest, because transmission is only one way. The major disadvantage of secondary radar is that it requires the installation of expensive electronic equipment in the aircraft, which must be 'friendly'. We shall mention this type of radar again at the end of section 3.2, and cover it more fully in chapter 6.

2.2 Statistical aspects of radar detection

In this section we introduce some important aspects of radar statistics, showing how they affect performance. Before getting down to detail it is helpful to consider the pulse radar waveforms in figure 2.2. At the top are shown two pulses from a transmitted pulse sequence (note, however, that in practice each pulse would typically contain hundreds, or thousands, of radio-frequency cycles − not the four or five indicated). Adjacent pulses are separated by the interpulse period, equal to the reciprocal of the PRF. The centre waveform shows a typical swept gain (or STC) characteristic.

The lower part of the figure shows a typical received signal after the application of swept gain. The broken lines correspond to breakthrough accompanying each transmitter pulse. Note that each echo should have many more cycles of oscillation, and that the signal's total dynamic range would normally be much greater than indicated.

In spite of these simplifications, the figure illustrates several important points. Firstly, we see that the early part of the signal contains unwanted echoes, known as *clutter*. These are likely to come from buildings or trees, or from irregular terrain. Fortunately they, too, are substantially reduced by swept gain. It can be very difficult to detect wanted targets in clutter, even when they produce a doppler shift. However, we have shown one strong echo (target 1) which is clearly visible in the clutter.

Radar propagation at microwave frequencies is normally line-of-sight. The Earth's curvature therefore tends to restrict strong clutter returns to the initial part of each interpulse period. Of course, this may not apply in hilly or moun-

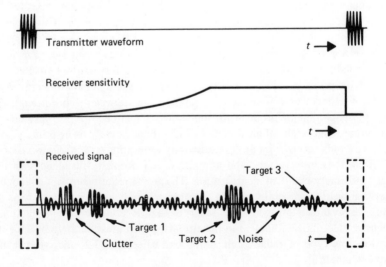

Figure 2.2 Pulse radar waveforms

tainous regions, or if the antenna is high above the ground. Nor is weather clutter confined to short ranges. Nevertheless, the figure is fairly typical.

Two additional target echoes are also shown. That from target 2 is strong and should be easy to detect. Target 3 is at long range and gives a much feebler echo. Although there is little clutter in this part of the received signal, weak echoes may well be masked by system noise. Much of this arises in the input stages of the receiver. We conclude that whereas clutter provides the main challenge to successful target detection at short range, long-range performance is generally limited by system noise.

The received signal is statistical in several important respects. Firstly, a given target does not always present the same effective cross-section to the radar. Small changes in target attitude can produce dramatic changes in the value of σ, and these are very unpredictable. Secondly, system noise is an essentially random phenomenon which can only be described in statistical terms. And finally, clutter is very difficult to define exactly. It is very site-dependent, and often varies with wind and weather. We now discuss target fluctuations and system noise in more detail, and consider their effect on the Radar Equation.

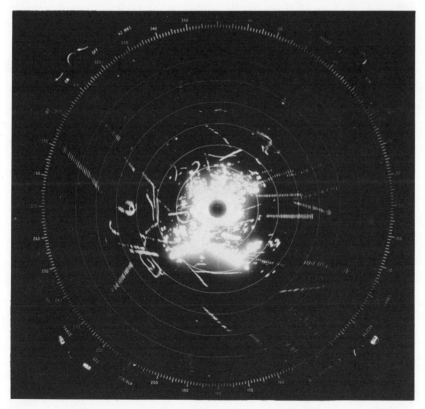

Figure 2.3 Typical clutter on the PPI display of a surveillance radar, prior to signal processing (photo courtesy of Marconi Radar Systems Ltd)

2.2.1 Target fluctuations

A complex radar target such as an aircraft has many reflecting surfaces – fuse-
lage, wings, tail and so on. Each of these contributes to the overall echo, taking
due account of relative phase as well as magnitude. As the target changes its
attitude with respect to the radar, constructive and destructive interference
between the various contributions can lead to large fluctuations in effective
cross-section. A complex target does not have a unique value of σ. We have to
work with some average, or representative, value.

Target fluctuations are very difficult to predict theoretically except in the
simplest of cases. With a real-life target we generally have to rely on practical
measurements – either on the target itself or on a scale model. The complexity
of the situation is well illustrated by figure 2.4. Part (a) shows a polar plot of
the composite echo produced by two reflectors placed one radio wavelength
apart. Each is taken as a point source, and assumed to re-radiate its incident
energy isotropically ($0°$ indicates that the line between the two sources is
normal to a line-of-sight from the radar). The scale rings on this figure are
10 dB apart. Even with this very elementary example of a 'composite target',
the pattern of re-radiation is quite complicated. Rotation would cause large
fluctuations in the effective value of σ.

Part (b) of the figure shows an equivalent plot for an aircraft, as it rotates
about a vertical axis. Scale rings are again drawn 10 dB apart. The details will
obviously depend on the size and type of aircraft, but σ is generally greatest
'sideways-on'. However, the most important conclusion is that small changes
in attitude can have dramatic effects on cross-section – often 20 dB or more.
Comparable results are obtained with ships and other complex targets.

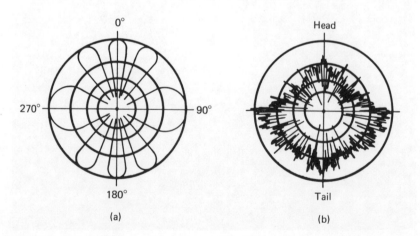

Figure 2.4 Polar plots of target fluctuations: (a) for a 'composite' target con-
sisting of two isotropic point-reflectors; and (b) for an aircraft

It is obviously difficult to decide on representative values of σ in such cases, for use in the Radar Equation. An average or median value is often used. The situation is further complicated by the fact that radar cross-sections depend to some extent on frequency. Fortunately, when predicting the performance of a radar system at the design stage, one often specifies a value of σ. This need not correspond to any particular practical target.

So far we have considered only the extent of target fluctuations. However, their time-scale is also important. For example, if an aircraft or ship were to maintain a strictly constant attitude with respect to a radar over a 1-minute period, a rotating antenna might get many opportunities for detection with a particular value of σ. In practice, moving targets rarely if ever behave in this way. Not only is σ likely to change between successive antenna scans, but in extreme cases it may even change significantly between successive transmitter pulses.

This is a good moment to mention that a radar antenna normally 'illuminates' a target with many transmitter pulses on each scan. The number of *hits-per-target* is typically between about 5 and 50. This has important implications for the build-up of each blip on a radar display, and for the design of MTI systems. We will discuss these matters later. For the moment it is sufficient to realise that a fast-fluctuating target may present different cross-sections to the radar on successive transmitter pulses, and hence within a single scan.

The radar researcher P. Swerling investigated such ideas quantitatively in the late 1950s. He proposed several alternative mathematical models to describe target fluctuation statistics. They may be summarised as follows.

Swerling case 1 Echo pulses from a target are assumed to be of constant amplitude on any one scan, but independent (uncorrelated) from one scan to the next. Scan-to-scan fluctuations are described by a probability density function (pdf) of the form

$$p(\sigma) = \frac{1}{\sigma_{av}} \exp\left(\frac{-\sigma}{\sigma_{av}}\right), \qquad \sigma > 0 \tag{2.7}$$

where σ_{av} represents the average target cross-section.

Swerling case 2 Echo pulses are assumed to be independent from pulse to pulse, as well as from scan to scan. This corresponds to a very rapidly fluctuating target. The pdf is again given by equation (2.7).

Swerling case 3 As case 1, except that the pdf is taken as

$$p(\sigma) = \frac{4\sigma}{\sigma_{av}^2} \exp\left(\frac{-2\sigma}{\sigma_{av}}\right), \qquad \sigma > 0 \tag{2.8}$$

Swerling case 4 As case 2, except that the pdf is given by equation (2.8).

Equation (2.7) is relevant to a complex target having many comparable echo areas. Aircraft normally come in this category, and the Swerling case 1 model is widely used in radar performance calculations. Equation (2.8) is more appropriate for targets having one dominant reflector, plus a number of other, smaller reflectors. If the completely non-fluctuating target (sometimes referred to as a case 0 or case 5 target) is included, we have a total of five possible target models.

It is found that target cross-section fluctuations generally reduce radar range performance. This is especially true for Swerling case 1 targets giving strong echoes and hence a high nominal probability of detection. The use of case 1 statistics in radar calculations therefore tends to produce substantially more conservative range estimates than are given by non-fluctuating targets. We shall have more to say about this important matter in the next section.

2.2.2 Receiver and system noise

We introduced the idea of a minimum detectable signal in section 2.1, and assumed a value of 10^{-13} W in our calculations. This is indeed of the right order for a typical radar receiver. But the actual value of S_{min} depends on a number of factors and choices which are intimately related to the statistics of radar detection.

One of the most important of these factors is *receiver noise*. Every communications receiver adds a certain amount of noise to its input signal, and a radar receiver is no exception. Even with very careful design, noise due to thermal motion of electrons in resistive components is unavoidable. The amount of such *thermal noise*, also known as *Johnson noise*, is proportional to receiver bandwidth. It equals kT_0B watts, where k is Boltzmann's constant, T_0 is the absolute temperature, and B is the bandwidth in hertz. At normal operating temperature ($T_0 = 290°$K), kT is about 4×10^{-21} watts/Hz.

The reader may be tempted to suggest bandwidth reduction as a solution to the problem of receiver noise. However, if the bandwidth is made too small the receiver does not amplify and process signal echoes properly. A compromise is required. In practice, the receiver bandwidth of a pulse radar is normally close to the reciprocal of the pulse duration. For example, a radar using 1 μs pulses may be expected to have a bandwidth of about 1 MHz. Bandwidth restriction is generally achieved by using a high-Q tuned IF amplifier (see figure 1.1).

Much effort has been devoted to the design of low-noise input stages for radar receivers. But however good the design, there is still a certain amount of noise generated in subsequent receiver stages, and in the input transmission line or waveguide. Further noise enters the system via the antenna.

For all these reasons the quantity kT_0B is regarded as a minimum input noise power, only attainable in an 'ideal' receiver. Noise in a practical receiver is invariably greater by some factor F_n, known as the *noise figure*. Thus the actual noise, referred to the receiver input, is

$$kT_0BF_n \text{ watts} \tag{2.9}$$

where T_0 is taken as 290°K. A noise figure is generally expressed in decibels, with a typical value between 2 and 10 dB. We may also think of noise figure as the factor by which a receiver degrades the signal-to-noise ratio of a signal passing through it.

Let us next suppose that successful detection of a target at its maximum range R_{max} is just possible if the input signal-to-noise power ratio is $(S/N)_1$. In this case the input signal power is, by definition, S_{min}. Hence

$$\left(\frac{S}{N}\right)_1 = \frac{S_{min}}{kT_0BF_n}, \quad \text{or} \quad S_{min} = kT_0BF_n\left(\frac{S}{N}\right)_1 \tag{2.10}$$

To develop this argument further we need an appropriate value for $(S/N)_1$. However, before proceeding it is important to note that we are talking about the detection of a target using a single transmitter pulse. Later on we will consider the effects of using more than one hit-per-target.

Although the signal-to-noise ratio $(S/N)_1$ is referred to the receiver input, we may assume a similar value at the output of the IF amplifier, immediately before detection. This is because almost all the noise arises in the initial stages of the receiver. The reader is perhaps tempted to suggest that any value of $(S/N)_1$ greater than unity will allow successful detection. However, it is more complicated than that! The basic reason is that noise is random. It cannot be assumed to behave in a particular manner at the instant when a target echo arrives. So we cannot expect to detect an echo with complete certainty — even if the signal-to-noise ratio is much greater than unity. All we can hope for is an acceptable probability of detection.

It is helpful to start by considering some important properties of noise. The noise voltage at the input to the IF amplifier is generally assumed to be *gaussian*, with zero mean value. A gaussian distribution is expected whenever noise arises as a result of many individual random events — such as the thermal motion of electrons in the input stages of a receiver. The probability density function (pdf) of gaussian noise with zero mean is given by

$$p(v) = \frac{1}{(2\pi\psi_0)^{1/2}} \exp\left(\frac{-v^2}{2\psi_0}\right) \tag{2.11}$$

where ψ_0 is the *variance* and $\psi_0^{1/2}$ is the *standard deviation (SD)*. Note that the variance is the same as the mean-square value, or AC power, of a waveform having zero mean.

A typical portion of gaussian noise with zero mean is shown at the top of figure 2.5a. It spends most of its time within about two standard deviations of the mean, with occasional high peaks. The probability of finding it within amplitude range v_1 to $(v_1 + dv_1)$ is simply equal to $p(v_1) dv_1$. In the lower part of figure 2.5a we show the gaussian distribution itself. This confirms that the noise is most likely to have a value close to zero, although there is a very small chance of finding it several standard deviations above or below the mean.

Like signal echoes, the noise passes through the IF amplifier and is then detected. As we have already explained, the IF amplifier is usually a high-Q bandpass amplifier which controls the overall bandwidth of the receiver. So the

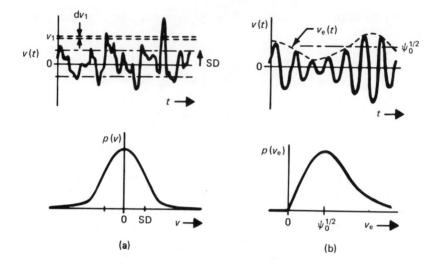

Figure 2.5 Noise characteristics

noise reaching the detector is narrowband. A typical portion of such noise is shown in part (b) of the figure. It looks rather like an amplitude-modulated sine wave, although it in fact contains a range of frequencies and has a random envelope. When the noise waveform is detected, the upper envelope $v_e(t)$ is extracted. It may be shown that the envelope has the so-called *Rayleigh* pdf

$$p(v_e) = \frac{v_e}{\psi_0^{1/2}} \exp\left(\frac{-v_e^2}{2\psi_0}\right) , \quad v_e > 0 \tag{2.12}$$

This is also illustrated in the figure. We see that the envelope spends much of its time close to $\psi_0^{1/2}$. However it, too, displays occasional large peaks.

The output from the detector is compared with a *threshold level*, to decide whether any target echoes are present. This may be shown to be the statistically optimum approach. Even so, there is unfortunately a danger that the noise will itself exceed the threshold, giving rise to a *false alarm*. This is shown by figure 2.6. The time-scale of the envelope $v_e(t)$ has been considerably compressed, compared with figure 2.5b. But, as before, it fluctuates around the value $\psi_0^{1/2}$, with occasional high peaks. One of these peaks exceeds the threshold voltage V_T, causing a false alarm. In other words, a noise peak has been wrongly interpreted as a target echo.

We may, of course, reduce the probability of false alarms by raising the detection threshold. But we are then less likely to detect genuine targets. This calls for a compromise. Too low a value of V_T gives many false alarms; but too high a value causes *misses* on targets, reducing the probability of successful detection. The reader will by now appreciate that we cannot sensibly talk about target detection without also considering false alarms.

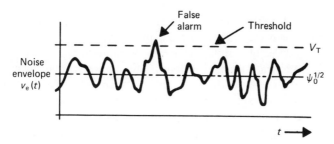

Figure 2.6 The generation of a radar false alarm

The probability of finding a random waveform above some threshold level V_T equals the area of its pdf between V_T and ∞. Hence the probability of obtaining a false alarm due to noise is given by

$$P_{fa} = \int_{V_T}^{\infty} \frac{v_e}{\psi_0^{1/2}} \exp\left(\frac{-v_e^2}{2\psi_0}\right) dv_e = \exp\left(\frac{-V_T^2}{2\psi_0}\right) \tag{2.13}$$

Now the average duration of each noise 'pulse' in figure 2.6 is approximately equal to the reciprocal of the bandwidth B. So the maximum rate of false alarms which could possibly occur (given a very low threshold) is about B per second. This information, together with equation (2.12), may be used to obtain the graphs shown in figure 2.7. Here the average time beween false alarms is plotted as a function of threshold level, for three values of receiver bandwidth. To make the graphs relevant to any values of V_T and ψ_0, we plot $V_T^2/2\psi_0$ rather than V_T itself. And we express it as a power ratio in decibels.

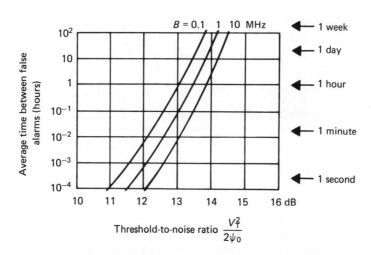

Figure 2.7 Average time between false alarms as a function of threshold setting and receiver bandwidth

How often can a false alarm be tolerated in a practical radar system? The answer must depend very much on the application. The acceptable false-alarm rate in a strategic early-warning radar may be quite different from that in a system designed to track weather balloons. Note, however, that a wide range of false alarm rates is encompassed by a narrow range of threshold values. In many practical cases we may expect $V_T^2/2\psi_0$ to be set between about 11 and 14 dB.

The false-alarm probabilities corresponding to acceptable false-alarm rates are very small. For example, an average of one false alarm per minute in a receiver of bandwidth 1 MHz corresponds to $P_{fa} = 1.7 \times 10^{-8}$. This implies that, in practice, the detection threshold must be set well into the extreme 'tail' of the Rayleigh distribution shown in figure 2.5b.

We have considered the question of false alarms with some care. We now need to consider what happens when a signal echo occurs, mixed with noise. There are two possibilities. Either the detected signal-plus-noise waveform exceeds the threshold, in which case we register a successful detection. This is called a *hit*. Alternatively, the waveform fails to reach the threshold, giving a *miss*.

The probabilities associated with these three types of occurrence – false alarm, hit, and miss – are closely interrelated. This is shown in part (a) of figure 2.8. The pdf for the noise alone is a Rayleigh distribution, as discussed already. The small, heavily hatched area to the right of the threshold represents the probability of a false alarm, P_{fa}. When a signal echo is superimposed on the noise, the range of possible amplitude values alters. The composite waveform is much more likely to exceed the threshold. A typical pdf for signal-plus-noise is also shown in the figure. The large, lightly hatched area to the *right* of the threshold gives the probability P_d of a correct detection, or hit. And the area

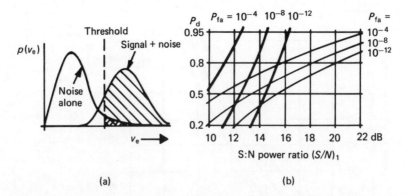

(a)   (b)

Figure 2.8 Statistics of false alarms, hits, and misses. (a) Typical probability density functions, and (b) relationships between detection probability, false-alarm probability and receiver signal-to-noise ratio. The heavy curves refer to a non-fluctuating target, the light curves to a fluctuating target with Swerling case 1 characteristics

under the same curve to the *left* of the threshold represents the probability of a miss.

All these probabilities depend, of course, on the position of the threshold, and on the strength of the echo signal compared with the noise. They also depend on the assumed target fluctuation statistics, because these affect the form of the signal-plus-noise pdf.

If a non-fluctuating target is assumed, the signal-plus-noise pdf takes the so-called Rician form described by the mathematician S. O. Rice in the 1940s

$$p(v_e) = \frac{v_e}{\psi_0} \exp\left(-\frac{v_e^2 + A^2}{2\psi_0}\right) I_0\left(\frac{v_e A}{\psi_0}\right) \tag{2.14}$$

Here A represents the amplitude of a sinusoidal target echo at IF frequency. I_0 is the modified Bessel function of zero order. As we would expect, this pdf reduces to the Rayleigh form for noise alone if $A = 0$ (see equation (2.12)). The situation is more difficult to visualise if the target fluctuates randomly. However, it may be shown that the commonly assumed Swerling case 1 target gives rise to a signal-plus-noise pdf which is also of the Rayleigh type.

The overall effects of these factors on the detection of a single pulse are summarised by figure 2.8b. The thick curves refer to a non-fluctuating target; the thin ones to a Swerling case 1 target. For an acceptable false alarm probability P_{fa}, and a desired detection probability P_d, we may read off the signal-to-noise power ratio $(S/N)_1$ required in the receiver. The figure shows just how troublesome target fluctuations can be — especially when high values of P_d are required.

We now illustrate the above discussion with an example. Suppose we have a receiver of bandwidth 1 MHz, and can tolerate a false-alarm probability of 10^{-6}. This corresponds to an average time between false alarms of 1 second. (The reader may feel that this is rather a high false-alarm rate. However it is important to realise that, since a valid target normally produces many hits per scan, a single false alarm is most unlikely to be interpreted as a target.) Let us further assume that we require a detection probability of 80 per cent on a Swerling case 1 target. The receiver noise figure is 5 dB, and the standard deviation of the noise immediately prior to detection is 100 mV. We wish to find

> The detector threshold level,
> The signal to noise ratio required at the receiver input,
> The minimum detectable signal S_{min}.

Figure 2.7 shows that an average false-alarm interval of 1 second in a receiver of 1 MHz bandwidth corresponds to

$$10 \log_{10}\left(\frac{V_T^2}{2\psi_0}\right) = 11.7 \text{ dB}, \quad \text{giving } V_T = 5.44 \ \psi_0^{1/2} \tag{2.15}$$

The noise standard deviation ($\psi_0^{1/2}$) is here equal to 100 mV. Therefore the threshold should be set to 544 mV.

Referring to figure 2.8b we see that a false-alarm probability of 10^{-6}, together with a detection probability of 0.8 on a Swerling case 1 target, requires a signal-to-noise ratio of about 17 dB.

The minimum detectable signal S_{min} may now be found using equation (2.10). 5 dB corresponds to a power ratio of 3.16, and 17 dB to a power ratio of 50.1. Hence

$$S_{min} = (kT_0)BF_n\left(\frac{S}{N}\right)_1 = 4 \times 10^{-21} \times 10^6 \times 3.16 \times 50.1$$

$$= 6.33 \times 10^{-13} \text{ W} \tag{2.16}$$

The value of S_{min} is greater than the 10^{-13} W assumed in section 2.1, implying a less sensitive receiver. However, one further important factor needs to be taken into account. This is the number of hits-per-target.

We have previously mentioned that a typical radar illuminates each target with between about 5 and 50 individual pulses on each scan. As the antenna sweeps through a strong target, a number of individual hits are obtained in rapid succession. In the simplest types of radar system, these hits are simply built up on the phosphor of the display, forming a composite 'blip'. With a small or distant target, a perfectly adequate blip may be obtained, even though not all individual echoes register as hits. This means that the detection probability on each scan is likely to be considerably better than the detection probability for an individual transmitter pulse.

The build-up of a composite blip on a radar display is called *video integration*. In more expensive and sophisticated systems, such integration is often performed by special electronic circuits. This may be done either before or after detection. The two techniques are referred to as *predetection*, or coherent, integration; and *postdetection*, or non-coherent, integration. The second is simpler to achieve, although less efficient.

Suppose a radar operates with n hits-per-target. Then it may be shown that an ideal predetection integrator would achieve an improvement in signal-to-noise power ratio of n times, compared with a single pulse. Postdetection integration (including that obtained using the screen phosphor) achieves considerably less. We may allow for non-ideal performance by defining an integration efficiency $E_i(n)$. Thus for n hits-per-target the actual signal-to-noise ratio improvement, compared with a single pulse, is

$$nE_i(n) \tag{2.17}$$

With a typical value of $n = 20$ and postdetection integration, $E_i(n)$ might be around 50 per cent. This would given an effective number of hits-per-target of about 10.

When a radar operator looks at a display, he is interested in detecting a target each time the antenna scans through it. He does not mind whether or not each individual transmitter pulse scores a hit. We may therefore reduce the value of

S_{min} given in equation (2.16), to take account of the beneficial effects of video integration. Thus

$$S_{min} = \frac{kT_0 BF_n}{nE_i(n)} \left(\frac{S}{N}\right)_1 \tag{2.18}$$

As a matter of interest, if we assume $n = 20$ and $E_i(n) = 0.5$, the numerical value of S_{min} in equation (2.16) changes to rather less than 10^{-13} W. This refers, of course, to a detection probability of 80 per cent on each scan, rather than each pulse.

Video integration also affects the significance of false alarms. Equation (2.13) derived the false-alarm probability P_{fa} on the assumption that approximately B independent decisions are made each second. P_{fa} is, in effect, a probability *per-pulse*. However, if the system operates with n hits-per-target, the operator waits for n echoes to be integrated before making a decision about the presence or absence of a target. The average decision rate is therefore reduced to B/n per second, and the false-alarm probability *per-decision* is $(n P_{fa})$. This larger value is sometimes quoted in radar literature. It is important to be clear which of the two definitions of false-alarm probability is being adopted. Here we always use the per-pulse value, P_{fa}.

Our discussion of the detection process has assumed a conventional, analog radar display, viewed by an operator who decides whether or not targets are present. In this situation the detection threshold V_T (see figure 2.6) may be thought of as the signal level which just causes visible 'bright-up' of the display. This is a rather vague notion. It depends on such factors as ambient lighting and the operator's skill and alertness. In modern systems there is an increasing trend towards *automatic detection*. Decisions about whether targets are present or not are made by electronic circuits or computers. The results of these decisions are then presented to the radar operator on a digital display. We shall say more about this approach in sections 5.7 and 7.4.

2.2.3 Other system factors

Various additional factors affect radar system performance. Most of them are not 'statistical' in the true sense of the word. They do not display short-term random fluctuations, like target cross-section or receiver noise. On the other hand some of them may be hard to determine precisely, and they may vary somewhat unpredictably in the medium or long term. For this reason it is convenient to include them in our discussion of radar statistics.

The factors fall broadly into three categories. These are design losses, operational losses and propagation losses. We now give a brief account of each.

By *design losses* we mean those factors which can be taken into account at the design stage, but which may be overlooked. Among the most important are the following.

Plumbing loss. Small power losses occur in the transmission lines or waveguides connecting the transmitter and receiver to the antenna, and in the duplexer. In a typical installation these may amount to a few decibels.

Antenna beam-shape loss. In our earlier discussion of hits-per-target and video integration, we implied that all echoes received from a non-fluctuating target are of equal amplitude. But, in practice, the train of received pulses is amplitude-modulated by the shape of the antenna beam. The antenna gain is effectively less for pulses at the beam edges. This may be allowed for by including a beam-shape loss factor in the Radar Equation. A typical figure for a surveillance radar is 1.6 dB.

MTI loss. We have previously noted that the use of MTI tends to produce blind speeds. This is equivalent to a loss of receiver sensitivity on targets having certain radial velocities. We will cover such losses more fully in section 5.4.

Collapsing loss. A target echo on a radar display must compete with unwanted clutter and system noise. If the resolution of the display is too coarse, clutter and noise from adjacent 'resolution cells' tend to interfere with echo visibility. Similar effects may be produced by inadequate receiver bandwidth. They are major causes of the complicated effect known as collapsing loss, which may amount to a few decibels.

Our second category comprises *operational losses*. When a radar system is installed − perhaps in a remote location − its performance is often somewhat inferior to that suggested by initial design calculations. The factors involved include equipment ageing, errors in equipment settings and maintenance problems. It is obviously very difficult to quantify these accurately, although they may easily account for a few extra decibels of loss in the system. Another operational factor is fatigue or stress in the radar operator.

Propagation losses make up the third category. They too are difficult to quantify. The main factors are as follows.

Atmospheric effects. We normally assume free-space propagation in radar calculations. However, the Earth's atmosphere has three main effects. These are attenuation, reflection and refraction.

Attenuation of radar waves at the normal radar frequencies is quite small except in heavy rainfall, and may usually be neglected. *Reflection* is much more important. As we have already noted in section 1.2.2, it can give rise to strong weather echoes − particularly at the higher radar frequencies.

Refraction is produced by the decreasing density of the atmosphere with height. It causes slight bending of radar waves, giving an increase in line-of-sight. This may be approximated by assuming the Earth's effective radius to be rather greater than its actual radius. A factor of 4/3 is commonly used.

Antenna lobe structure. When a radar beam is directed at low elevation angles, some of its energy almost inevitably illuminates the ground. Apart from the ground clutter this produces, energy reaching a target comprises both direct and ground-reflected components. Constructive or destructive interference between these components can cause either a reinforcement (lobe) or a partial cancellation (null). The resulting lobe structure can have an important effect on radar coverage. It will be discussed more fully in section 3.2.

It is obviously difficult to include most of the above factors directly in the Radar Equation. Many of them are strongly influenced by the particular site or application, or by the weather. However, this makes them no less real or important. One approach is to include an additional factor L_s in the equation to cover 'system losses', accepting that it is unlikely to be very accurate. If this is not done, it should cause little surprise if the actual field performance of a radar system falls short of theoretical predictions.

2.2.4 Effects on the Radar Equation

We developed a simple form of the Radar Equation at the beginning of this chapter. Equation (2.5) gave the maximum range on a target as

$$R_{max} = \left(\frac{P_t G \sigma A_e}{16\pi^2 S_{min}} \right)^{1/4}$$

(2.19)

For ease of reference we again define the various symbols:

P_t = transmitter power output
G = antenna gain
σ = effective target area
A_e = effective antenna area
S_{min} = minimum detectable signal

We have since shown that some of these parameters are more complicated than they seem, and that various additional factors and losses should be taken into account. Let us now derive a form of the Radar Equation which may be expected to give a more realistic prediction of system performance.

Equation (2.18) expressed the minimum detectable signal as

$$S_{min} = \frac{kT_0 B F_n}{nE_i(n)} \left(\frac{S}{N} \right)_1$$

(2.20)

where

k = Boltzmann's constant
T_0 = absolute temperature
$kT_0 \approx 4 \times 10^{-21}$ watts/Hz at normal operating temperatures
B = receiver bandwidth in hertz
F_n = receiver noise figure

$\left(\dfrac{S}{N} \right)_1$ = input signal-to-noise ratio

n = number of hits-per-target

$E_i(n)$ = integration efficiency

We showed that the signal-to-noise ratio required at the receiver input is a function of the false-alarm probability P_{fa}, and of the single-pulse detection probability P_d. Additionally, we have suggested in section 2.2.3 that a loss factor L_s be included in the Radar Equation to represent various system losses. L_s may be thought of as reducing the effective transmitter power output. With this modification, and substituting for S_{min} in equation (2.19), we obtain:

$$R_{max} = \left[\frac{P_t}{L_s} \times \frac{G\sigma A_e n E_i(n)}{16\pi^2 k T_0 B F_n \left(\dfrac{S}{N} \right)_1} \right]^{1/4} \qquad (2.21)$$

So far we have assumed a transmitter power output P_t, and that some specifiable signal-to-noise ratio is required at the receiver input. But we have taken no account of the type of transmission used – CW, FMCW, pulse or pulse-doppler. In fact, if we use average transmitter power rather than peak pulse power in the equation, it becomes relevant to a wide variety of radar systems. This assumes that *matched-filter* detection is used in the receiver – a topic to be covered in section 5.1. In the case of a pulse radar, we may write the average transmitter power as

$$P_{av} = P_t f_p \tau, \quad \text{giving} \quad \frac{P_{av}}{P_t} = f_p \tau \qquad (2.22)$$

where

f_p = pulse repetition frequency (PRF)

τ = pulse length

The ratio of average power to peak pulse power is known as the *duty cycle* of a pulse radar. It equals the proportion of the total time for which the transmitter is actually turned on. In a typical system the duty cycle may be around 0.001. In other words, the average transmitter power output is only about one-thousandth of the peak power output.

Another helpful modification is to express the effective antenna area as

$$A_e = A\rho_A \qquad (2.23)$$

where

A = actual physical area of antenna

ρ_A = antenna efficiency

ρ_A is less than unity, and represents the imperfect collection of incident echoes by a practical antenna. Incorporating equations (2.22) and (2.23) into equation (2.21), we obtain

$$R_{max} = \left[\frac{P_{av}}{f_p \tau L_s} \times \frac{G\sigma A \rho_A n E_i(n)}{16\pi^2 k T_0 BF_n \left(\frac{S}{N}\right)_1} \right]^{1/4}$$

or

$$R_{max}^4 = \frac{P_{av} G A \rho_A \sigma n E_i(n)}{16\pi^2 k T_0 F_n B \tau f_p \left(\frac{S}{N}\right)_1 L_s} \tag{2.24}$$

Although this version of the Radar Equation is considerably more sophisticated than equation (2.19), it is by no means the only possible one. Other operating parameters could well be included, or substituted. To mention just two examples: the number of hits-per-target (n) could be expressed in terms of the PRF, antenna beamwidth and antenna rotation speed; and the antenna gain (G) could be expressed in terms of its area and the radio wavelength. The reader may therefore expect to find alternative versions of the equation in different books and references.

Two further comments should be made about equation (2.24). Note firstly that the calculated range assumes free-space propagation and a target which appears as a point-source to the radar. (In other words the angle subtended by the target at the radar is assumed negligible compared with the antenna beamwidth.) Secondly, the effects of weather and clutter on target detectability are ignored — unless some attempt is made to include them using the system loss factor L_s.

In spite of these reservations and limitations, the equation may be expected to give realistic predictions of performance. As an example, consider a small marine radar of the type used on sailing boats and other small vessels. Suppose we need to know the maximum range at which it can detect a non-fluctuating target of effective area 0.1 m^2. A typical specification might include the following parameters:

$$P_t = 3 \text{ kW} \quad G = 25 \text{ dB} \quad A = 0.08 \text{ m}^2 \quad \rho_A = 0.6$$

$$F_n = 10 \text{ dB} \quad B = 4 \text{ MHz} \quad \tau = 0.25 \text{ } \mu s \quad f_p = 3 \text{ kHz}$$

We will use the following additional values:

$$E_i(n) = 0.5 \quad k T_0 = 4 \times 10^{-21} \text{ W Hz}^{-1} \quad L_s = 8 \text{ dB}$$

Note that in this system $P_{av} = P_t f_p \tau = 2.25 \text{ W}$. We will also assume that the antenna rotates at 40 rpm and has a horizontal beamwidth (in plan view) of $3.5°$. The beam therefore takes $(1.5 \times 3.5/360) = 0.0146 \text{ s}$ to scan through a point target. At the PRF of 3 kHz this gives $n = 44$ hits-per-target.

We must also specify the signal-to-noise ratio $(S/N)_1$. Let us assume a false-alarm rate of one every 10 seconds. Since the receiver bandwidth is 4 MHz, this gives a false-alarm probability of about $(10 \times 4 \times 10^6)^{-1} = 2.5 \times 10^{-8}$. Suppose also that we require a detection probability on each pulse of 50 per cent. Refer-ring back to figure 2.8, we see that for $P_{fa} = 2.5 \times 10^{-8}$ and $P_d = 0.5$ we need an input signal-to-noise ratio of about 12.4 dB. Thus, finally, we have

$$P_{av} = 2.25 \text{ W} \qquad n = 44 \qquad \left(\frac{S}{N}\right)_1 = 12.4 \text{ dB}$$

The reader may like to check that the use of these values in equation (2.24) gives a maximum range on a 0.1 m^2 target of about 2.5 km, or 1.35 nm.

Even if our range prediction should prove realistic, we must beware of placing too much faith in the calculations. There are always uncertainties surrounding the operation of a practical radar system. Perhaps more valuable than the precise value produced by equation (2.24) is the chance it offers to compare radar systems with one another, and to assess the effects of changing system para-meters.

3 Operational and Siting Factors

The successful installation of a radar system depends on a number of operational and siting factors. A key one is the reduction of clutter. In part, this may be achieved by careful siting of the antenna. Another important item is energy reflection towards targets from a land or sea surface. As noted in section 2.2.3, this may substantially alter the antenna beam pattern. This chapter begins by describing some basic properties of clutter, and goes on to discuss radar coverage and siting in air-traffic-control and marine applications.

3.1 Radar clutter

The clutter problem has been mentioned several times in previous chapters, and we have shown an example of clutter in figure 2.3. The term *clutter* describes any unwanted echoes on a radar display. These may come from birds, buildings, trees, terrain, sea-waves or weather. Of course, echoes which constitute clutter in one application may represent useful signals in another. A good example is weather echoes. They are generally a nuisance in air-traffic-control applications, but certainly not in radar meteorology.

There are various ways of reducing radar clutter using signal processing techniques. An important one is swept gain, or STC. As we pointed out in section 2.1, this technique is widely used to reduce receiver sensitivity on short-range targets. Since the Earth's curvature tends to confine clutter echoes to the early part of each interpulse period, STC also has a major effect on clutter. However, it does not improve the signal-to-clutter *ratio*. Separate techniques must be used to achieve this. Some of them are discussed in chapter 5. The present section concentrates on the basic nature of clutter, and the ways in which it is affected by antenna beamshape and siting.

Clutter is hard to quantify. In many ground-based systems it varies dramatically with *azimuth* (that is, bearing from the radar site). For example, there may be a city in one direction, mountains in another. The clutter seen by a marine radar depends markedly on sea-state and wind direction. Weather clutter is inherently variable and unpredictable. In spite of these difficulties, it is important to have an overall appreciation of clutter generation and statistics.

We first consider the generation of *surface clutter*. This is illustrated by figure 3.1. The upper part of the figure shows a radar antenna mounted above a land

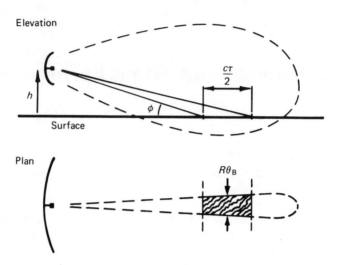

Figure 3.1 The generation of surface clutter

or sea surface. It is assumed to point more-or-less horizontally, and some of its energy strikes the surface. (We have indicated a typical vertical beamshape by a broken line, however its precise form is unimportant here.) The corresponding plan view is also shown. Note that in most cases the horizontal beamwidth is much smaller than the vertical beamwidth, to give good resolution in azimuth. This is achieved by making the antenna's width considerably greater than its height.

Surface clutter is often distributed over a considerable area, unlike the point targets we have discussed previously. Its backscattering effect may be described in terms of a radar *cross-section density*, σ_0. If a clutter area A_c produces an effective radar cross-section σ_c, we write

$$\sigma_0 = \frac{\sigma_c}{A_c}, \quad \text{or} \quad \sigma_c = \sigma_0 A_c \tag{3.1}$$

Let us now consider energy striking the surface at an angle ϕ. This is called the *grazing angle*. It is helpful to specify a clutter area corresponding to one *resolution cell* of the radar. Such a cell may be thought of as equivalent to the smallest distinguishable area on the radar display. Echoes occupying the same resolution cell cannot be separated. In terms of the figure, we see that the cell's extent in the range direction is determined by the pulse length. A pulse of duration τ corresponds to a go-and-return path of $c\tau/2$. The equivalent distance along the clutter surface is $c\tau \sec\phi/2$. The cell's extent in azimuth equals the product of the range R and the horizontal beamwidth θ_B. The surface area lying within one resolution cell is therefore

$$A_c = \frac{c\tau \sec\phi}{2} (R\theta_B) \tag{3.2}$$

In effect the radar 'lumps together' all individual clutter echoes received from this area. Any wanted target occupying the same resolution cell must, of course, compete with the composite clutter echo.

The effective radar cross-section of the clutter area is

$$\sigma_c = \sigma_0 A_c = \sigma_0 \frac{c\tau \sec\phi}{2} (R\theta_B) \tag{3.3}$$

Equation (2.4) gave an expression for the echo power P_r received from a target of area σ

$$P_r = \frac{P_t G\sigma A_e}{16\pi^2 R^4} \tag{3.4}$$

Let us now suppose that a point target of area σ competes with clutter of effective area σ_c in the same resolution cell. We denote their respective echo powers by S and C. It is clear from equation (3.4) that the signal-to-clutter power ratio must be

$$\frac{S}{C} = \frac{\sigma}{\sigma_c} = \frac{2\sigma}{\sigma_0 c\tau \sec\phi \, R\theta_B} \tag{3.5}$$

If the clutter is severe, the clutter power will be much greater than the receiver noise power. Hence the maximum range at which a target is detectable depends on the signal-to-clutter ratio, rather than the signal-to-noise ratio. If the minimum acceptable signal-to-clutter ratio at the receiver input is $(S/C)_{min}$, then

$$\left(\frac{S}{C}\right)_{min} = \frac{2\sigma}{\sigma_0 c\tau \sec\phi \, R_{max}\theta_B} \tag{3.6}$$

We conclude that the Radar Equation is very different when system performance is limited by surface clutter, rather than receiver noise. For example, an increase in the transmitter power P_t does not improve target detectability. Stronger target echoes are simply offset by stronger clutter echoes. Equation (3.6) also suggests a maximum range proportional to target area — unlike equation (2.5), which gave a fourth-root relationship. This implies that maximum range is much more dependent on target area in the clutter-dominated case. We should also note that an effective way of improving the signal-to-clutter ratio is to reduce the size of the radar resolution cell. This may be achieved by reducing the pulse length, the antenna beamwidth, or both.

Equation (3.6) should be applied and interpreted with caution. It assumes an even distribution of clutter over a large horizontal surface. Such conditions are very unlikely to be encountered in practice, even in marine environments. We should rather expect σ_0 to vary widely as a function of range and azimuth. The equation is nevertheless valuable for suggesting the major influence of surface clutter on the Radar Equation.

What are the actual values of the clutter cross-section density σ_0? In practice this parameter is found to depend heavily on the type of surface, or terrain, and on the grazing angle ϕ. It is also affected by the choice of radar frequency, and by the *polarisation*. The term polarisation refers to the direction of the electric field vector in the radiated energy. *Horizontal* and *vertical* polarisation are both widely used in radar systems. *Circular* polarisation gives a useful reduction of echoes from spherical rain droplets. This is important in medium-range and long-range systems operating at 10 cm or less, which tend to suffer from severe rain clutter. However, circular polarisation is more difficult and expensive to achieve. The detailed variation of σ_0 with all these factors is highly complicated. But if we confine ourselves to the small grazing angles ($\phi < 10°$) which are of most interest in air-traffic-control and marine radar, we may use table 3.1.

Table 3.1

Type of surface or terrain	σ_0 (dB)			
	L-band		X-band	
	Horiz.	Vert.	Horiz.	Vert.
City	−15	−15	−18	−15
Cultivated land	−30	−25	−25	−22
Sea	−45	−35	−40	−30

It must be emphasised straight away that table 3.1 is very approximate. Actual values depend heavily on the particular site. The quoted sea clutter values are typical of 'average' sea conditions, in wind strengths of around 15 knots. We note that clutter echoes are generally more severe from city buildings than from the sea, with cultivated land somewhere in between. σ_0 is generally slightly larger for vertical polarisation, and at higher radar frequencies. It also tends to increase substantially for grazing angles above about $10°$.

Surface clutter is important in many radar applications, but not all clutter is of this type. Individual buildings, radio masts, water towers, birds, and so on, are likely to appear as point clutter. Their effect on performance may be assessed in a similar way to point targets.

A third type of clutter is *volume clutter*, caused by rain or other atmospheric conditions. It may also be produced by deliberate scattering of metallic foil strips, or *chaff*, in an attempt to confuse a radar. Volume clutter is usually quantified in terms of an effective cross-section η per unit volume. Thus, if a clutter volume V_c produces an effective radar cross-section σ_c, we write

$$\eta = \frac{\sigma_c}{V_c}, \text{ or } \sigma_c = \eta V_c \tag{3.7}$$

This compares directly with equation (3.1). Note, however, that a radar resolution cell now corresponds to a volume in space rather than a surface area. Its extent is defined in range by the pulse length, in azimuth by the horizontal beamwidth θ_B, and in elevation by the vertical beamwidth ϕ_B. The approximate volume of a resolution cell is therefore

$$V_c = \frac{c\tau}{2} (R\theta_B)(R\phi_B) \tag{3.8}$$

In practice, the most important type of volume clutter comes from rain and cloud droplets. They are normally very small compared with the radar wavelength, giving rise to Rayleigh scattering. Under these conditions the cross-section presented by an individual droplet is proportional to the sixth power of its diameter. Heavy rain produces stronger clutter than light rain or cloud, not only because there are more droplets per unit volume, but also because they tend to be larger. Droplet cross-section is also proportional to the fourth power of the transmitter frequency. This means that systems operating at the lower radar frequencies are much less susceptible to weather clutter — a matter already discussed in section 1.2.2.

Various empirical relationships are used to describe the effects of rain clutter on radar performance. One of these is

$$\eta \approx 7 f_0^4 r^{1.6} \times 10^{-12} \, \text{m}^{-1} \tag{3.9}$$

where f_0 is the transmitter frequency in GHz and r is the rainfall rate in mm/h. As an illustration, let us compare the cross-sections presented by light rain ($r = 1$) for two different radar wavelengths (23 cm and 3 cm), at a range of 50 nm. We will assume a pulse length $\tau = 2 \, \mu$s, a horizontal beamwidth $\theta_B = 1.5°$, and a vertical beamwidth $\phi_B = 12°$. The reader may like to verify that these values give a resolution cell volume of about 1.1×10^{10} m^3. Using equations (3.7) and (3.9), we find that the effective radar cross-section is about 0.2 m^2 at $\lambda = 23$ cm and about 800 m^2 at $\lambda = 3$ cm. Of course, in practice rain is most unlikely to fill a resolution cell completely (for one thing, it never falls from infinite height!). So actual cross-sections would be smaller. Nevertheless, these figures graphically illustrate the difference in weather performance between 23 cm and 3 cm radars. They also help explain why S-band and L-band frequencies are so often chosen for medium-range and long-range surveillance applications.

Before leaving the topic of volume clutter, we should mention the fascinating topic of *angels*. This is the term given to radar echoes for which no obvious reflecting objects exist. Angel echoes caused great consternation in the early days of radar. However, it is now known that they are produced by birds, insects or atmospheric inhomogeneities. Birds flying in formation may generate quite extensive angels – often in arc or ring patterns. Quite small concentrations of insects can also cause serious clutter on a radar display. Atmospheric angels are produced by variations of refractive index, and are associated with various types of clear-air turbulence.

Our next task is to consider *clutter statistics*. In the discussion of receiver noise in section 2.2.2, we noted that random noise arising as the superposition of many statistically independent processes or events has a gaussian amplitude distribution. This is also true of a clutter echo, provided it is made up of contributions from many independent scatterers within each resolution cell. In other words, amplitude fluctuations in the clutter echo from cell to cell are described by the gaussian form of probability density function (pdf) given in equation (2.11). After IF amplification and detection, fluctuations in the *clutter envelope* conform to the Rayleigh distribution (see also equation (2.12))

$$p(v_e) = \frac{v_e}{\psi_0^{1/2}} \exp\left(\frac{-v_e^2}{2\psi_0}\right), \quad v_e > 0 \tag{3.10}$$

Clutter with this form of pdf is known as *Rayleigh clutter*.

The condition for Rayleigh clutter – namely, many independent scatterers within each resolution cell – is commonly met in the case of weather echoes, and echoes from smooth, homogeneous terrain. The Rayleigh model also applies to sea clutter if the sea is calm and the resolution cell is fairly large. But in a short pulse-length, high-definition, marine radar, the size of individual wave features is often comparable with a resolution cell. This is more likely to happen with rough seas. The effect is to increase the probability of high peaks, and hence false alarms, caused by the sea clutter. In practice, therefore, the detection threshold of a marine radar is often set rather higher than would be expected on the basis of equation (3.10).

Various alternative forms of pdf are used to quantify sea clutter in marine radars. Their common feature is a somewhat longer 'tail' than the Rayleigh distribution (see figure 2.5b). One of these is the *Weibull* family of distributions, of which the Rayleigh is a limiting case. After envelope detection, a Weibull distribution gives a pdf of the form

$$p(v_e) = \alpha \ln 2 \left(\frac{v_e}{\beta}\right)^{\alpha-1} \exp\left(-\ln 2 \left[\frac{v_e}{\beta}\right]^\alpha\right), \quad v_e > 0 \tag{3.11}$$

where α and β are constants. An alternative is the *log-normal* distribution, giving an envelope pdf

$$p(v_e) = \frac{\gamma}{v_e} \exp\left(-\delta\left\{ \ln\left[\frac{v_e}{\beta}\right] \right\}^2 \right), \quad v_e > 0 \tag{3.12}$$

where β, γ and δ are constants. The choice of distribution depends on various factors, of which sea-state is the most important. The log-normal function, which produces the longest 'tail', is best suited to rough seas. Weibull and log-normal distributions have also been quite widely used to model clutter from 'awkward' land surfaces, such as forests, mountainsides and cities.

We end this section with a brief discussion of *clutter movement*. Of course, objects such as buildings or bare mountainsides are fixed with respect to a ground-based radar. Clutter echoes received from them contain no doppler shift. However, moving objects – birds, trees and crops swaying in the wind, sea-waves, rain droplets and chaff – do, in general, produce a doppler shift. They may be interpreted as 'moving targets' by a signal processing system designed to detect aircraft, and allowed to obscure the radar display. It is clearly important to have some idea of the frequency shifts produced by typical moving clutter.

Like many other aspects of radar, this is a complicated question. All we can do here is reproduce a few empirical results. Figure 3.2 shows some typical power spectra of moving clutter for a transmitter frequency $f_0 = 1$ GHz ($\lambda = 30$ cm). Each curve has been normalised to its zero-frequency value, and therefore indicates relative power. The shapes of the curves are reasonably well fitted by functions of the form

$$W(f) = \exp\left(-a\left[\frac{f}{f_0}\right]^2\right) \approx \exp\left(\frac{-f^2\lambda^2}{8\sigma_v^2}\right) \tag{3.13}$$

Here a is a dimensionless constant, and σ_v is the *rms velocity spread* of the clutter in m s^{-1}. Both depend upon the type of clutter. They are related by

$$\sigma_v \approx \frac{c}{(8a)^{1/2}} \tag{3.14}$$

The caption to figure 3.2 identifies the type of clutter and the value of σ_v relevant to each curve.

A simplified explanation of the relationship between the frequency scale of figure 3.2 and the values of σ_v is as follows. Equation (1.1) gave the doppler frequency of a moving target with radial velocity v_r as

$$f_d = 2v_r \frac{f_0}{c} \tag{3.15}$$

If, for example, $v_r = 1$ m s^{-1} and $f_0 = 1$ GHz, then $f_d = 6.67$ Hz. Of course, the individual scatterers in moving clutter have a complex statistical distribution of radial velocities, rather than one particular value. It is nevertheless interesting to note that (for example) curve (d) in the figure, for which $\sigma_v = 1$ m s^{-1}, reduces to around half its initial value in the region of 7 Hz. We may therefore think of σ_v as representing an approximate, average radial velocity of the moving clutter. It is also worth noting that doppler shift is proportional to f_0 in equation (3.15). Therefore the curves in figure 3.2 should be reasonably valid for other transmitter frequencies, provided the abscissa is scaled in proportion. These results will be useful for our discussion of MTI systems in chapter 5.

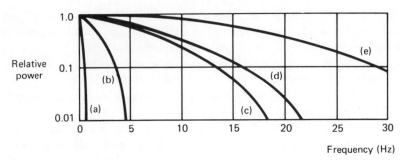

Figure 3.2 Typical power spectra of moving clutter at a transmitter frequency of 1 GHz:
  (a) lightly wooded hills, calm day (σ_v = 0.02 m s^{-1});
  (b) heavily wooded hills, windy day (σ_v = 0.2 m s^{-1});
  (c) sea-surface, windy day (σ_v = 0.9 m s^{-1});
  (d) chaff (σ_v = 1 m s^{-1});
  (e) rain clouds (σ_v = 1.9 m s^{-1}) [adapted from Barlow (1949) – details in entry 7 in the Bibliography at the end of this book]

3.2 Horizontal and vertical coverage

A radar antenna normally produces either a *pencil beam* or a *fan beam*. A pencil beam has approximately equal horizontal and vertical beamwidths, giving axial symmetry. It is widely used for tracking applications – for example, in radars designed to follow missiles. A fan beam generally has a narrow horizontal beam-width, but a much wider vertical beamwidth. Target resolution is good in azimuth, but not in elevation. The effect is to illuminate any target on a given bearing from the radar, regardless of its height above the ground. This is appro-priate in surveillance and search radars which produce a 'bird's eye view' of an area on a plan position indicator (PPI) display.

Our development of the Radar Equation in chapter 2 included the antenna gain parameter G. It defines the power gain of the antenna at the centre of the beam, compared with an isotropic radiator. In practice, of course, there is a distribution of power in both azimuth and elevation. It is often conveniently summarised in terms of the 'half-power' beamwidths θ_B and ϕ_B (see equations (3.2) and (3.8)). However, these do not tell the whole story. They merely specify the angular intervals over which the radiated power density is within 50 per cent of its maximum value. A proper discussion of horizontal and vertical coverage requires a rather fuller description of antenna beamshape.

Although we reserve detailed comments about antenna design until section 4.2, some further general points should be made here. Firstly, an important basic principle of antenna theory is *reciprocity*. This states that the properties of an antenna are identical on transmission and reception. It follows that the effective receiving area A_e and the gain G are proportional to one another. The actual relationship is

$$A_e = A\rho_A = \frac{G\lambda^2}{4\pi} \qquad (3.16)$$

where A is the physical area of the antenna and ρ_A is its efficiency. We concentrate here on the radiation pattern during transmission; but the principle of reciprocity shows that we could equally discuss coverage in terms of the gain pattern during reception.

There is also a link between the antenna parameters G, θ_B and ϕ_B. The precise relationship depends upon the particular antenna, but a useful approximation is

$$G \approx \frac{20\,000}{\theta_B \, \phi_B} \qquad (3.17)$$

where the beamwidths are expressed in degrees. Thus we would expect a typical surveillance antenna with $\theta_B = 1.5°$ and $\phi_B = 12°$ to have a maximum power gain of around 1110, or 30.5 dB.

It might seem desirable to make a radar beam as narrow as possible. By concentrating the radiated energy, we achieve good target resolution and enhance the range. Furthermore, as shown in the previous section, the use of a small resolution cell tends to reduce the problem of clutter echoes. However, there are other considerations. If the resolution cell is very small, it takes a long time for the radar to search the required volume. There is said to be a low *data rate*. This can be improved by speeding up the antenna, but individual targets then receive fewer hits-per-scan, giving a lower probability of detection. A compromise is needed — and this does not always result in a very narrow beam.

Another important aspect of radar coverage concerns the *sidelobe performance* of the antenna. It is not possible, in practice, to focus all the radiated energy into a single main lobe. Figure 3.3 shows a polar plot of the horizontal radiation pattern for a typical surveillance radar antenna. For convenience, the scale rings are drawn at -10 dB intervals with respect to the main lobe. The main lobe is narrow. However, there are a number of additional lobes, referred to as *sidelobes*, *spillover radiation* and *back lobe*. These are undesirable for several reasons. Firstly, strong echoes from nearby targets or clutter may enter the receiver via the subsidiary lobes, degrading or confusing the display. The system is also much more susceptible to interference from nearby radars, or to hostile *jamming*. Finally, power directed into subsidiary lobes is wasted. It would be better used in the main beam.

The figure suggests a typical first sidelobe level about 25 dB down on the main lobe. There are additional, smaller sidelobes. The spillover radiation at about $90°$ to the main lobe is caused by energy 'escaping sideways' from the antenna's reflecting surface. There is also, in general, a significant back lobe. All of these can be reduced by careful design and manufacture, and at additional cost. We shall have more to say about sidelobe performance in section 4.2.

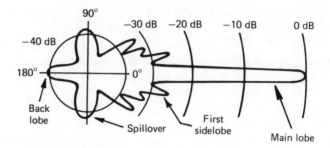

Figure 3.3   Horizontal radiation pattern of a radar antenna

We next discuss coverage in the vertical plane. There are two main issues. Firstly, we need to consider the basic vertical coverage provided by an antenna. This is referred to as its *free-space pattern*, or its pattern *in-the-clear*. We also need to assess the likely modifications to this pattern caused by any ground or sea reflections.

In the case of ground-based radar for air traffic control, it is often desirable to obtain echo signals which are independent of range for constant-altitude targets. The vertical cover provided by a simple fan beam does not achieve this, because too little power is radiated towards aircraft at high altitude close to the radar. The antenna must therefore be modified to radiate more strongly at the higher elevation angles, approximating the so-called *cosecant-squared* pattern.

Consider an aircraft target at elevation angle ϕ, range R and height h. It is simple to deduce that

$$\frac{h}{R} = \sin \phi, \text{ or } R = h \operatorname{cosec} \phi \tag{3.18}$$

Referring back to equation (2.4), we see that the power received from a target at range R may be expressed as

$$P_r = \frac{P_t G \sigma A_e}{16 \pi^2 R^4} \tag{3.19}$$

Equation (3.16) gives the proportional relationship between A_e and G. Substituting for A_e and R we obtain

$$P_r = \frac{G^2 P_t \sigma \lambda^2}{64 \pi^3 h^4 \operatorname{cosec}^4 \phi} \tag{3.20}$$

If the height h is constant, and P_r is to be constant regardless of elevation angle, it follows that

$$\frac{G^2}{\operatorname{cosec}^4 \phi} = \text{constant, or } G \propto \operatorname{cosec}^2 \phi \tag{3.21}$$

Hence the gain of the antenna in the vertical plane should be proportional to the cosecant-squared of the elevation angle. Practical antennas are often designed to approximate this pattern. It is not possible to do so over the full range of elevation angles, but cosecant-squared shaping might typically be attempted between $\phi = 5°$ and $\phi = 30°$. Note that this type of pattern is also appropriate for airborne radars which map the Earth's surface.

There is one important qualification to the cosecant-squared idea. We have pointed out in sections 2.1 and 3.1 that swept gain, or sensitivity time control (STC), is often used to reduce strong echoes from nearby targets and clutter. However, it also reduces echoes from close aircraft at high elevation angles. To compensate, the antenna pattern must be adjusted to provide even higher gain at these elevations than the cosecant-squared pattern.

We should not assume that cosecant-squared vertical coverage is always ideal, or even desirable. Marine radars designed to detect coastlines and targets on the sea surface do not need it. The antennas of such radars often have a very small vertical aperture (that is, height), giving a wide but crudely shaped vertical beam. However, they save on size, weight, wind resistance and cost. Another example is precision approach radars (PARs) used in air traffic control. These do not generally need to follow high-flying aircraft at short range.

Let us now consider the question of reflections from the ground or sea surface, first mentioned in section 2.2.3. Such reflections tend to cause lobes and nulls in the vertical coverage pattern. The geometry of the situation is illustrated by figure 3.4. The antenna (A) is mounted at height h_a above a horizontal reflecting surface. A target (T) is at height h_t and elevation angle ϕ with respect to the radar. The *path length* of a ray passing directly from antenna to target equals the range R. This is the normal range used in radar calculations, referred to as the *slant range*. The path length of a ray reflected from the surface is R'. We assume the angles of incidence and reflection are equal. The total energy arriving at the target may be found by superimposing the direct and reflected rays, in phase as well as amplitude. If they add constructively, there will be a lobe; if destructively, a null.

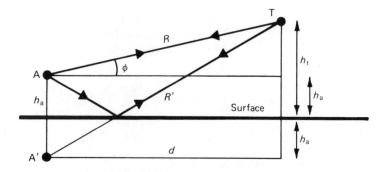

Figure 3.4 The geometry of surface reflections

To assess this situation quantitatively, we need to find the *path difference*, and hence the phase difference, between reflected and direct rays. A convenient approach is to represent the effect of the reflecting surface by a virtual, mirror-image antenna (A'). Noting that the horizontal distance to the target is d, we have

$$R^2 = d^2 + (h_t - h_a)^2 \text{ and } (R')^2 = d^2 + (h_t + h_a)^2 \qquad (3.22)$$

Furthermore

$$(R')^2 - R^2 = (R' + R)(R' - R) \approx 2R(R' - R) \qquad (3.23)$$

The path difference is therefore given by

$$(R' - R) \approx \frac{(R')^2 - R^2}{2R} = \frac{d^2 + (h_t + h_a)^2 - [d^2 + (h_t - h_a)^2]}{2R}$$

$$= \frac{2h_a h_t}{R} \qquad (3.24)$$

Now

$$\frac{(h_t - h_a)}{R} = \sin \phi$$

hence

$$(R' - R) \approx \frac{2h_a h_t}{(h_t - h_a)} \sin \phi \qquad (3.25)$$

Let us assume that, for aircraft targets, $h_t \gg h_a$. Then, to a very good approximation, the path difference is

$$(R' - R) \approx 2h_a \sin \phi \qquad (3.26)$$

The corresponding phase difference is

$$\psi_d \approx 2h_a \sin \phi \, \frac{2\pi}{\lambda} \text{ radians} \qquad (3.27)$$

If we consider low elevation angles for which $\sin \phi \approx \phi$, then

$$\psi_d \approx \frac{4\pi h_a \phi}{\lambda} \text{ radians}$$

The indirect ray often undergoes an additional phase shift of approximately π on reflection from the surface (as discussed below). Hence the total phase difference is given by

$$\psi_d + \pi = \frac{4\pi h_a \phi}{\lambda} + \pi \text{ radians} \qquad (3.28)$$

When this equals $n\pi$, n being an even integer, there is constructive interference, giving a lobe in the vertical coverage. This occurs when

$$\frac{4\pi h_a \phi}{\lambda} = (n-1)\pi, \ n \ \text{even} \tag{3.29}$$

The lowest of these lobes corresponds to $n = 2$, giving

$$\frac{4\pi h_a \phi}{\lambda} = \pi, \ \text{or} \ \phi = \frac{\lambda}{4h_a} \tag{3.30}$$

By similar arguments, if n is an odd integer, we get a null. The first of these occurs at $\phi = 0$, the next at $\phi = \lambda/2h_a$, and so on.

Figure 3.5 shows a coverage diagram with a lobe pattern. It refers to a surveillance radar operating at 23 cm wavelength (L-band), with an antenna height of 15 m. Note that the different scales used for height and range produce a distortion of the elevation angles. Clearly, the coverage shown must refer to a particular set of parameters — transmitter power, target size, probability of detection and so on — but we do not need to consider these here. Using equation (3.29) we may deduce that adjacent lobes (or nulls) are spaced by 0.0077 radian, or 0.44°. The coverage pattern in the clear is shown by a broken line. In effect, the lobes and nulls caused by ground reflections represent modulations of this pattern. Since the lobe structure continues up to about $\phi = 6°$, we infer that the lower cut-off of the free-space pattern must be at about $\phi = -6°$ in this case. The maximum range in the clear is obtained at an elevation angle of about 2°, implying that the antenna's axis is tilted up slightly from the horizontal. Note also that this particular antenna provides roughly cosecant-squared coverage between about $\phi = 4°$ and $\phi = 30°$.

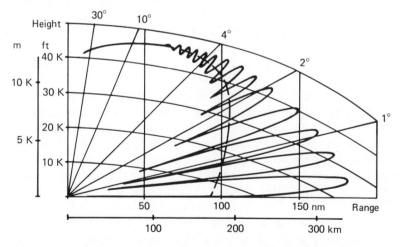

Figure 3.5 Typical vertical coverage pattern for a surveillance radar, showing the lobes and nulls caused by ground reflections

We have so far explained the angular separation of the various lobes and gaps, but not their magnitudes. Clearly, these must depend on the amount of power radiated on to the horizontal surface by the antenna. The second major influence is the proportion of incident power reflected from the surface towards the target. This is expressed in terms of a *reflection coefficient* ρ. The value of ρ is affected by the grazing angle, the roughness of the surface, and to a lesser extent by its electrical properties. It also depends on the radar frequency. However, when horizontal polarisation is used, the magnitude of ρ is normally very close to unity, with a phase angle of π (as already assumed in equation (3.28)). So horizontal polarisation tends to produce a well-developed lobe-null structure, as in figure 3.5. Vertical polarisation is more complicated. The reflection coefficient varies widely with grazing angle, and its amplitude passes through a minimum at the so-called *Brewster angle*. Generally, the overall effect is to give less-pronounced lobes and nulls than horizontal polarisation. Circular polarisation is intermediate in this respect.

It must be stressed that figures 3.4 and 3.5 are based on a very simple model of surface reflections. Essentially, we have used the elementary laws of geometrical optics to infer the positions of lobes and nulls. In practice, the reflection of energy from a land or sea surface is partly *specular*, partly *diffuse*. The specular component behaves in a 'geometrical' way, and predominates if the surface is very smooth. It is characterised by coherent phase relationships between contributions from adjacent points on the surface. The coefficient ρ applies to this component. The diffuse component, on the other hand, represents disorganised omni-directional scattering, with incoherent phase relationships. It is difficult to include such effects in theoretical predictions of vertical coverage patterns.

Even if we ignore the above complications, there is another important factor affecting the generation of lobes and nulls. Real land and sea surfaces are never perfectly 'flat and infinite'. The antenna of a ground-based radar sees different terrain features and undulations at different bearings. The effective height of the antenna above the ground may also vary as the antenna rotates. Even a calm sea-surface is subject to the Earth's curvature. For these reasons we should regard equations (3.29) and (3.30) as giving an approximate indication of the lobe pattern, when the site is reasonably flat and extensive. In other cases, the lobes and nulls are likely to break-up into complicated patterns which depend on azimuth. Furthermore, practical measurements generally show that the lobes are less pronounced, and the nulls less deep, than predicted by theory.

Before leaving the topic of vertical coverage, we should consider the measurement of aircraft altitude using radar. Air-traffic-control procedures typically demand a vertical separation between aircraft of 1000 feet (say 300 m). Military requirements for height measurement may be more stringent. Referring back to figure 3.5, we see that this represents a big problem in terms of resolution in elevation. For example, two aircraft at a range of 100 nm, on the same bearing but separated by 1000 feet in altitude, subtend an angle of less than 0.1° at the radar. It would require a vertical beamwidth of the same order to resolve them as separate targets, implying an antenna with a large vertical aperture. Height-finding primary radars are indeed made; but they tend to be cumbersome and expensive. In the civilian environment it is far better to 'ask' a co-operating aircraft how high it is flying, using secondary radar. This point was previously mentioned at the end of section 2.1, and will be developed properly in chapter 6.

3.3 Air-traffic-control and marine radar

We next discuss the practical siting of air-traffic-control (ATC) and marine radars. The ATC environment is often particularly difficult, for several reasons. Operational requirements and the problems posed by clutter tend to be more severe. There may also be a number of *possible* sites for the radar antenna. Although this gives welcome flexibility, it means that a range of choices and compromises have to be balanced against each other. By contrast, the positioning of a marine radar antenna is often dictated by the height of a vessel's masts and superstructure.

Starting with ATC radar, it is helpful to outline the needs of civil aviation for aircraft surveillance. Radar is only one of many electronic navigation aids. However, it is unique in providing controllers with a comprehensive view of air traffic over a wide area. It has come to assume a crucial role in modern ATC procedures, especially in busy airspace.

Central to ATC procedures is the concept of *controlled airspace*. Aircraft using this airspace must be suitably equipped, and their pilots properly qualified. We may divide controlled airspace into three parts:

Control zones, surrounding and protecting major airports

Terminal control areas, normally designated at the confluence of airways near major airports

Airways, connecting terminal areas and linking up with the airways of adjacent countries. Airways are corridors of airspace 10 nm wide, often from a base of about 6000 feet (say 1800 m) to a height of 24 500 feet (say 7500 m)

In the UK, special rules apply to *upper airspace* above 24 500 feet. This is used by en-route jet aircraft. All civil and military aircraft, many of which are over-flying, are subject to a mandatory ATC service, including radar control.

The ATC service ensures that aircraft are safely separated. Aircraft using airways under radar surveillance are kept at least 5 nm apart horizontally, or 1000 feet apart vertically. As an aircraft continues on its journey, it is handed over from one *flight information region (FIR)* to another. It finally makes contact with the destination airport. A terminal area or approach control radar is commonly used to monitor and advise during the approach phase, leading the aircraft towards the runway.

The various phases of ATC are not easily separable, but the problems are usually most serious near major airports. Figure 3.6 illustrates a fairly typical situation, for in-bound traffic only. Descending aircraft enter the terminal area (T) via several airways (A). During busy periods, ATC allocates each aircraft to the top of a holding pattern or *stack* (S). Aircraft are withdrawn in turn from the bottom of the stack, and the rest of the traffic ladders-down to fill the empty levels. On leaving the stack, an aircraft follows a prescribed route to the runway (R), establishing itself on the correct flight-path for landing.

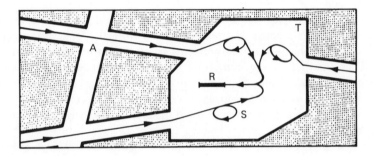

Figure 3.6 Controlled airspace in the vicinity of a major airport

The operational requirements for terminal area and approach control radars are therefore likely to be complicated. It may be necessary to monitor aircraft to within a short distance of touchdown. Slow light aircraft may be mixed in with jet traffic, with its different operating heights and rates of descent. When we consider that aircraft are also taking off, and some may be overflying, it is clear that the required radar coverage will depend on particular terminal area boundaries, airway patterns, stack locations, runway headings and so on. There are generally different requirements in different directions.

All these factors must be considered when siting the antenna. Placing it well above the surrounding terrain allows an uninterrupted view of air traffic. It also gives good coverage on low-flying aircraft some distance from the radar. Modifications to the vertical coverage caused by ground reflections tend to be slight. The main disadvantage of high positioning is increased land clutter. Good anti-clutter performance is generally essential.

In practice, such a radar is often sited inside an airport perimeter, with its antenna placed on a tower or airport building. However, if an airport lies in a valley it may be essential to move the antenna to high ground. The radar signals may be sent back to the air traffic controllers by landline or radio link.

Surveillance of en-route aircraft demands radars of greater range and power; but in other respects it is less demanding than the terminal area and approach

Figure 3.7 A 10 cm approach control radar. By increasing the transmitter power, this system may also be used for terminal area surveillance (photo courtesy of Marconi Radar Systems Ltd)

roles. The need to minimise weather clutter makes a long wavelength (say 23 cm or above) essential. Antennas therefore tend to be large. Their siting is often rather difficult from an environmental point of view. However, in other respects it is more flexible. There is an increasing tendency for such systems to be remote-controlled, sending their signals to a distant ATC centre for integration with information from other radar sites.

We now turn briefly to marine radar. As already noted, the positioning of an antenna on a ship or boat is often dictated by the heights of masts and super-structure. It is generally desirable to raise the antenna as high as possible, par-ticularly when the radar is required to detect targets on the sea surface at con-siderable range. Simple geometrical considerations show that the distance to the horizon for an antenna at height h above the sea surface is approximately

$$d = (Dh)^{\frac{1}{2}} \tag{3.31}$$

where D is the Earth's effective diameter. The refraction of radar waves by the Earth's atmosphere makes the radar horizon somewhat longer than the normal optical horizon. The effective diameter is often taken as 4/3 times the actual diameter for these purposes — although this is only an approximation. Equation (3.31) then becomes

$$d = 1.23\, h^{\frac{1}{2}} \ (d \text{ in nautical miles}, h \text{ in feet})$$

or

$$d = 4.1\, h^{\frac{1}{2}} \ \ (d \text{ in km}, h \text{ in metres}) \tag{3.32}$$

For example if $h = 5$ m (typical of a small boat), $d = 9.2$ km; if $h = 25$ m (typical of a large vessel), $d = 20.5$ km. This does not mean that surface targets are necessarily easy to detect out to such ranges. They have to compete with sea clutter. Furthermore, surface reflections tend to produce a partial null in radar coverage at very low elevation angles. Fortunately, the 3 cm wavelength typical of many marine radars, together with the antenna heights used, produce a very fine lobe-null structure, with a narrow bottom null.

Equations (3.31) and (3.32) emphasise the line-of-sight limitations of normal marine radar. Of course, if the radar target is well above the sea surface (for example, another vessel with a high superstructure; or a high coastline), it can be detected at a greater distance. In this case we may add the distance to horizon of the radar antenna to that of the target. Nevertheless, line-of-sight limitations can be a considerable embarrassment to naval vessels requiring protection from low-flying aircraft or missiles. An effective solution is to install early warning radar in helicopters or aircraft attached to the fleet.

The line-of-sight problem explains why normal marine radars operate with modest transmitter power, and why the maximum displayed range is rarely above 50 nm. In fact the more powerful marine radars are used as much for their ability to detect small targets at close range, as for their long-range performance.

Figure 3.8 Typical marine radar antennas on a naval vessel (photo courtesy of Racal Marine Radar Ltd)

4 Radar Hardware

4.1 Transmitters, receivers and duplexers

The function of the transmitter in a pulse radar is to produce high-energy pulses with the required waveform, radio frequency and repetition rate. The transmitter has two major subsystems. These are the transmitting device itself, and the pulse modulator which turns it on and off (see figure 1.1). We now describe them in turn.

Three types of transmitting device have been dominant in radar for many years: the *magnetron*, the *klystron* and the *travelling wave tube*. All are based on vacuum-tube technology. The magnetron oscillator, first mentioned in section 1.1, has been continuously improved since its invention in 1940. Magnetrons are relatively cheap and rugged, and very suitable for mobile equipments. The klystron amplifier, first used in radar in the 1950s, offers the system designer a rather higher peak power capacity than the magnetron. Since it is an amplifier rather than an oscillator, it can provide highly stable output pulses, controlled in phase as well as frequency. This is a great advantage for MTI systems. On the other hand, klystrons are rather bulky, expensive and prone to mechanical damage. The travelling wave tube, a close relation to the klystron, is also an amplifier. One of its most valuable features is a relatively large bandwidth. Each of these tubes has advantages and disadvantages. The choice between them depends on the particular radar application.

There is also much interest in solid-state devices for radar transmitters. Individual devices offer much lower power ratings than their vacuum-tube competitors. This disadvantage may be partially offset by combining many devices in a single transmitter — as in the *multiple transistor module*. Such modules are particularly suitable for use with *phased-array antennas*, which combine the power from individual transmitter elements in space. We shall have more to say about this in section 4.2.2. Solid-state transmitters are inherently rugged and reliable. They do not need the high operating voltages of vacuum tubes, with the attendant danger of electric shock and X-rays. It seems likely that the next ten years will see further significant developments in this area, including the direct incorporation of transmitter devices into radar antennas.

We now describe in general terms the operation of the three main types of vacuum tube. Figure 4.1a shows the classic magnetron in its basic form. A central cathode is surrounded by a substantial copper anode. The anode contains a number of holes and slots which act as *resonant cavities*. When the cathode is heated, its oxide-coated surface emits electrons. These tend to move towards the anode under the influence of the applied DC electric field. However, there is also a strong DC magnetic field, supplied by permanent magnets (not shown). This second field is perpendicular to the plane of the diagram. As the electrons gather speed in the *interaction space*, they are forced into curved paths by the magnetic field, and pass the mouths of the cavities. In doing so, they excite the cavities at their resonant frequency. Since all the cavities are coupled together by the electron stream, radio-frequency (RF) energy may be extracted from any one of them using an output coupling loop or slot leading into a waveguide.

Some electrons return to bombard the cathode, causing *secondary emission* and an increased cathode temperature. The cathode heater may often be turned off once the oscillations have started. Another important feature of magnetron operation is that an anode with N resonant cavities can support any one of $N/2$ possible modes of oscillation. The preferred mode is the *π-mode*, in which there is an RF phase inversion between adjacent cavities. This may be encouraged by connecting alternate segments of the anode block with conducting *straps*. Even so, the conventional magnetron is subject to *mode instability*.

Improved stability may be obtained by coupling alternate resonators by a co-axial cavity completely surrounding the anode. The resulting *co-axial magnetron* is shown diagrammatically in part (b) of the figure. Although more expensive than a conventional magnetron, the co-axial type gives better efficiency and a longer service life, in addition to improved mode control.

Various techniques have been devised for tuning magnetrons by inserting rods or pistons into the resonant cavities. Frequency changes up to about 6 or 8 per cent are commonly achieved. Rapid frequency variations are sometimes needed for *frequency-agile* radar — for example, to counter hostile jamming. These may be produced by a rotating disc or oscillating piston which alters the cavity geometry. However, such methods are rather awkward and, as we shall see a little later, klystrons or travelling wave tubes are preferable for frequency-agile systems.

High-power magnetrons can produce several megawatts of peak power, and several kilowatts of mean power. Anode-to-cathode voltages for such devices are typically tens of kilovolts. Since the conversion efficiency from DC to RF energy is generally less than 50 per cent, there may be substantial cooling problems. Magnetrons are compact, stocky devices which are relatively easy to house. A high-power device might typically be 25 or 30 cm across.

We now consider the second type of transmitter tube — the klystron. Klystrons may be designed as oscillators at low power levels, but the high-power types used in pulse radar are invariably amplifiers. As already mentioned, the resulting pulses can be made highly stable in phase as well as frequency. This is a great advantage for moving-target indication, in which doppler frequency shifts are detected as changes in RF phase. Such a system is said to be *coherent*.

Figure 4.2a shows a simplified view of a high-power klystron. A heated *cathode (Ca)* acts as a source of electrons. These are formed into a narrow beam

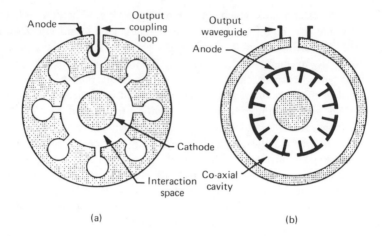

(a)   (b)

◀ Figure 4.1   (a) The classic pulse magnetron, and (b) the co-axial magnetron.
The photograph is of a high-power magnetron with a typical peak
pulse output of 2.3 MW at about 1.3 GHz. Anode and cathode
assemblies are shown alongside. The device operates with an anode
voltage of about 40 kV, a peak anode current of about 150 A, and
pulse duration of 5 μs. The overall weight, including magnet, is
86 kg, and the size is substantially increased by the need for elaborate
cooling arrangements (photo courtesy of English Electric Valve
Company Ltd)

and accelerated towards the *anode (A)* by the electron gun. They then pass
along the long *drift-tube* towards the *collector (C)*. During their journey along
the tube they are prevented from dispersing by an axial magnetic field, produced by coils or permanent magnets (not shown).

Arranged along the drift-tube are several resonant cavities, with interaction
slots opening into the tube. As the electrons pass the input cavity, they are
velocity-modulated by the input RF signal. In other words, some of them are
speeded up and others are slowed down, according to the instantaneous phase
of the RF signal. The interaction causes bunching and spreading of the electron
stream further down the tube. By placing an output cavity at the point of maximum bunching, large amounts of RF power may be extracted. Most high-power
klystrons have between one and three intermediate resonant cavities to increase
the bunching effect, and hence the power gain. Gains between 40 and 60 dB are
typical. On reaching the far end of the drift-tube, the electrons are removed by
a collector. Most of the waste heat appears at this electrode. When used as a
pulse radar transmitter, the electron beam is turned on and off between pulses
to reduce heat dissipation and increase efficiency.

The bandwidth of a klystron depends on its resonant cavities. If all are tuned
to the same centre frequency (*synchronously tuned*), the bandwidth is small and
the gain is high. However, the cavities are more commonly *stagger-tuned*, giving a
typical bandwidth of a few per cent, with a lower gain. Rather like the magnetron, the klystron centre frequency may be adjusted by altering the resonant
cavity geometry. Moving cavity walls or tuning paddles are often used, giving
tuning ranges up to about 15 or 20 per cent. Since the klystron is an amplifier,
frequency-agility is much easier to achieve than with a magnetron. It is only
necessary to vary the frequency of the relatively low-level drive signal — ensuring,
of course, that the klystron's bandwidth is not exceeded.

Very high pulse powers of 20 MW or more are obtainable. In part, this is
because the electron gun, interaction space and collector are well separated.
Each can be designed for optimum performance. However, klystrons are long
and rather awkward to house. DC voltages in excess of 100 kV may be required,
causing X-ray hazard.

The travelling wave tube (TWT) has many similarities with the klystron. It,
too, is an amplifier and works on the linear-beam principle. However, in the

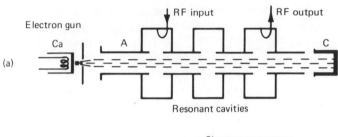

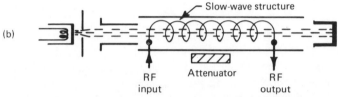

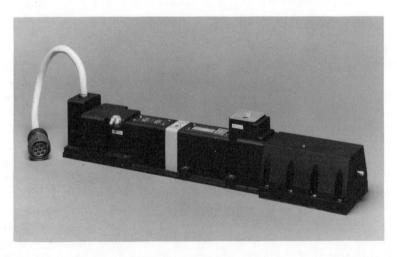

Figure 4.2 Basic geometry of (a) the drift-tube klystron, and (b) the travelling
wave tube (TWT). The photograph shows a medium-power TWT
mounted in its permanent-magnet focusing system. The device
operates at about 10 GHz, with a peak power output of 850 W, and
pulse length of up to 60 μs (photo courtesy of English Electric
Valve Company Ltd)

TWT the interaction between the electron beam and the RF field occurs over
the entire length of a *slow-wave structure*. This is illustrated in figure 4.2b. Once
again, velocity-modulation causes bunching and spreading of the electron beam
as it passes along the tube. An axial magnetic field must also be provided to keep
the beam focused. The input RF signal enters via a coupler or waveguide, and
propagates along the slow-wave structure. Although we have shown a simple

helix in the figure, modern high-power tubes use a variety of structures. Their essential task is to slow down the propagation velocity of the RF wave nearly to that of the electron beam. Effective interaction is then possible. RF output power is removed from the far end of the tube.

An important difference between the TWT and the klystron is that feedback is possible along the slow-wave structure of a TWT. This may lead to undesirable oscillation. The problem is overcome by incorporating a directional attenuator to reduce any backward wave, without significantly affecting the forward power transfer.

The gain, efficiency and power capability of a TWT are generally somewhat less than in a klystron of similar size. Even so, peak pulse powers of several megawatts are possible. An important feature of the TWT is its greater bandwidth — typically 10 to 20 per cent of the centre frequency. This is of particular value for radar systems requiring good range resolution, obtained by use of very short pulses or modern pulse compression techniques. It also makes frequency-agility relatively easy to achieve.

One further advantage of pulse compression systems should be mentioned here. In such a system each transmitter pulse is deliberately 'stretched' in time — typically by a factor of between 50 and 500 times. Received echoes are sharpened up using special signal processing techniques. This means that the *peak* power output from the transmitting device can be greatly reduced, while maintaining the *mean* power. Peak electrical stresses are correspondingly lower, giving improved device reliability and service life, with less electrical and X-ray hazard. Of course, this is of most value in long-range, high power radars.

It is now time to consider the second major component of a pulse radar transmitter — the *modulator*. This subsystem has the task of turning the transmitting device on and off. A simplified block diagram of a conventional modulator is shown in figure 4.4. During each interpulse period the electronic switch S is open, and the DC supply charges up an energy store via an inductive charging impedance. When a pulse is required, the switch is closed. Stored energy is rapidly discharged into the RF device via a pulse transformer. The charging impedance prevents significant pulse energy from feeding back into the DC supply. The energy store may take the form of a *LC* delay line or *pulse-forming network*. This has the useful property of generating a rectangular pulse. The modulator is then referred to as a *line-type modulator*. Essentially, the switch controls the start of the pulse, but the delay line controls the pulse length and trailing edge. For many years gas-filled tubes such as hydrogen thyratrons were used as modulator switches. However, these have now been superseded by solid-state devices.

A disadvantage of the line-type modulator is that the trailing edge of the pulse tends to be rather ragged. It is also difficult to alter pulse shape or duration. An alternative approach is to use an *active-switch modulator*, in which the switch controls both leading and trailing edges of the pulse. Modern solid-state switches allow most of the modulator to operate at relatively low voltage levels,

Figure 4.3 Transmitter–receiver cabinets and travelling wave tube of the
Watchman S-band surveillance radar. Note the air-cooling ducts for
the TWT, and the waveguide run to the antenna. The TWT power
output is about 60 kW peak, 1.2 kW mean. A number of advanced
features are incorporated, including pulse compression (photo
courtesy of Plessey Radar Ltd)

reducing hazard and electrical stress on components. The pulse transformer steps
the voltage level up to that required by the transmitter tube.

It is important to realise that a radar transmitter contains far more than a
modulator and a power output device. It requires power supplies, monitoring
equipment, protection and safety devices, and cooling systems. It also needs
elaborate timing and control electronics. When a klystron, TWT or solid-state
transmitting device is used, signal generators and drive amplifiers are required.
Many of these subsystems are as complicated as the high-power devices we have
described.

Before leaving the topic of radar transmitters, we should mention *frequency
diversity* operation. Our discussion so far has assumed the use of a single transmitter, with or without frequency agility. However, practical systems quite often

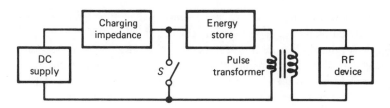

Figure 4.4   A radar pulse modulator

have two transmitters. One way of working them is to connect one to the antenna, keeping the other on standby in case of failure. However, there are considerable advantages in operating both transmitters simultaneously, but on slightly different frequencies. If the pulse train from one transmitter is delayed by a few pulse lengths, both can work into the same antenna. Dual receivers and signal processors are provided, the echoes being re-aligned before display. This is a full *dual-diversity system*.

An obvious advantage of the arrangement is improved reliability, since one channel can continue working even if the other fails. This is sometimes referred to as a *fail-soft* capability. There is also a doubling of the system's mean power output. According to the Radar Equation (see equation (2.5)) we would expect range performance to improve by a factor of $2^{0.25}$, or 19 per cent. In practice, the overall effect turns out rather better than this, because of the statistical nature of radar targets. As discussed in section 2.2.1, effective target area fluc-tuates widely with small changes in target attitude. However, such effects depend on radio frequency. On average, it is unlikely that fading due to target fluctuations will occur simultaneously on both channels of a frequency diversity system. (We assume that the two frequencies used are significantly different from one another — say at least a few per cent.) The net effect depends on the target fluctuation statistics, and on the required detection probability. For Swerling case 1 targets and $P_d = 80$ per cent, the range benefits of dual-diversity operation are typically about 25 per cent.

An alternative approach to frequency diversity involves working a single transmitter at two or more distinct frequencies. For example, the relatively large bandwidth of a TWT allows frequency changing every few pulses, or even pulse-to-pulse. The method is closely related to frequency agility, as discussed earlier. Of course, it does not produce the range benefits of increased mean power. Nor does it offer the reliability of a full dual system. But it is consider-ably cheaper.

As a complete contrast, we now consider the low-power end of a radar system — the receiver. We confine ourselves to some general remarks, accepting that the design of a solid-state radar receiver is a highly specialised business. We should start by noting that there is no clear division between the receiver and the signal processing circuits which follow it. Both make an essential contribution to the successful detection of signal echoes in noise and clutter. However, we focus here on the initial part of the processing chain, as far as the IF amplifier.

We have already described the problem of receiver noise in section 2.2.2, showing that it forms a fundamental limitation to detection of distant targets. A good noise figure is therefore one of the major requirements of a radar receiver, especially in a high performance system. Values of around 2 dB are attainable with modern low-noise 'front-ends', often based upon FET amplifiers. It is worth remembering that a reduction in receiver noise has the same effect on range performance as a corresponding increase in transmitter power. Except in systems where interference from nearby radars or hostile jamming is a problem, improving the noise figure may well be the better and cheaper option.

Another special requirement of radar receivers is the ability to handle signals with a very large dynamic range. Suppose, for example, a terminal area radar is required to detect a light aircraft with an effective area of 1 m^2, at ranges up to 100 nm. It should also cover large aircraft with an area of 20 m^2 to within 1 nm of the antenna, which is sited near the runway threshold. The Radar Equation (see equation (2.5)) tells us that the echo powers from these two targets are expected to be in the ratio $20 \times (100)^4$, or 93 dB. We should also remember that clutter close to the radar often provides even stronger echoes than large aircraft.

An important technique for making good use of the receiver's dynamic range is swept gain, or sensitivity time control (STC). It has already been discussed in sections 2.1 and 3.1. STC is often applied as a time-dependent attenuation at the input port of the receiver. It is valuable in allowing the receiver to handle echoes from nearby targets and clutter, and in reducing unwanted echoes from small targets such as birds. However, it does not improve the signal-to-clutter ratio. This last point may be illustrated by supposing that a signal echo arrives in the receiver at the same instant as an unwanted clutter echo 60 dB greater. This is quite a realistic assumption. In terms of voltage amplitude, the clutter echo is 1000 times stronger. Whatever clutter rejection circuits are used later in the processing chain, they cannot work if the receiver is unable to distinguish between the clutter echo on its own, and the clutter echo with the smaller signal echo superimposed. This is almost certain to happen if the clutter echo causes severe limiting in the receiver's input stages. A small superimposed signal then makes no difference to the receiver output.

One way of reducing the problem is to use a *logarithmic receiver*. This has an output voltage amplitude more or less proportional to the logarithm of the input signal level. (Actually, the characteristic cannot continue right down to zero input level. The receiver is generally linear at low levels, tending to logarithmic at high levels.) The logarithmic law is normally designed into the IF amplifier stage. In effect, the technique compresses the total amplitude range of signal and clutter echoes, avoiding receiver saturation. It does not, however, distinguish between signals and clutter. Additional circuits are needed for this purpose.

Referring back to figure 1.1, we recall that the typical pulse radar receiver works on the superheterodyne principle. It includes a low-noise RF amplifier, followed by a mixer and local oscillator to translate the frequency down to IF. The IF stage has a typical centre frequency in the range 30 to 70 MHz, with a narrow bandpass characteristic. It determines the overall bandwidth of the receiver. When the echoes are conventional rectangular pulses, the bandwidth which optimises the signal-to-noise ratio is approximately equal to the reciprocal of the pulse length. More generally, however, the IF amplifier should have the characteristics of a *matched filter*. This is a particularly important concept in the case of pulse compression systems. We should also emphasise that the receiver chain shown in figure 1.1 is very basic. It has no facilities for detecting the doppler shifts of moving targets, nor does it include any clutter suppression. All these topics will be covered in the next chapter.

We have previously noted that waveguides are generally used to connect the transmitter and receiver to the antenna. A waveguide is a hollow metal tube, normally air-filled and with a rectangular or circular cross-section. It should be made from highly conductive material such as brass, aluminium or copper. The inner surface must be smooth, and the cross-section uniform. A waveguide can exhibit various modes of propagation, with different electric and magnetic field patterns. Rectangular waveguides are widely used for microwave radar, with propagation in the *fundamental transverse-electric* (TE_{01}) mode. This mode allows transmission of wavelengths less than $2d$, where d is the wider dimension of the guide. Wavelengths up to about 30 cm may therefore be transmitted in guides of reasonable physical size. At the point where the waveguide run meets the antenna, a *rotating joint* must be incorporated. In many designs, this involves a transition from rectangular to circular waveguide. The whole waveguide system must be manufactured and assembled with great care, to minimise losses caused by junctions, discontinuities and surface imperfections.

When a single antenna serves both transmitter and receiver, a *duplexer* must be included in the waveguide system. As already mentioned, its role is to channel transmitter power to the antenna, and incoming echoes to the receiver. It must also protect the receiver from excessive transmitter breakthrough. Various types of duplexer have been developed. All rely on *transmit-receive (TR) cells* in one form or another. The conventional TR cell is a gas discharge tube, which acts as an electronic switch. Whenever the transmitter turns on, the cell breaks down and provides a short-circuit. This may be used to create a path from transmitter to antenna, and/or to protect the receiver by short-circuiting its input terminals. When the transmitter pulse ends, TR cell recovery allows incoming echoes to be routed to the receiver. Duplexer design involves the strategic placing of TR cells within the waveguide system to achieve the desired switching and protection.

Unfortunately, gas discharge tubes take some time to switch on, and even longer to recover. They must generally be 'primed', and are sometimes activated by a control pre-pulse. For these reasons, fast solid-state switches such as PIN diodes are increasingly specified. When used for receiver protection, these can also provide swept gain, or STC, as described in sections 2.1 and 3.1. Isolation between transmitter and receiver can be further improved by the use of non-reciprocal ferrite devices.

4.2 Antennas

We have already made a number of references to the properties of radar antennas. For example, we have introduced the notions of gain, effective receiving area and efficiency (equations (2.2) and (2.23)). In chapter 3 we discussed horizontal and vertical beamwidths in relation to the size of the radar resolution cell. The idea of reciprocity was introduced, and we gave approximate formulae relating various antenna parameters (equations (3.16) and (3.17)). Typical coverage patterns were shown in figures 3.3 and 3.5. We also considered antenna side-lobes, cosecant-squared vertical cover, and the lobe patterns produced by ground-reflected energy. More generally, it has been emphasised that long-range systems need wavelengths of 10 cm or above, to give reasonable freedom from weather clutter. This leads to relatively large antennas.

All the above ideas are general. In the present section we describe radar antennas in rather more detail, concentrating on two major types which have found widespread practical application.

4.2.1 Parabolic reflectors

The majority of radar antennas use some form of parabolic reflector to produce a narrow beam. The principle is straightforward, and well known in optics. Figure 4.5a shows that if an energy source is placed at the focus F of a parabola, then all energy striking the surface is reflected parallel to the axis. Furthermore, the path length from F to a plane AB lying perpendicular to the axis is the same for all reflected rays. This means that a point source at F is converted into a plane wavefront with uniform phase.

If the parabola is rotated about its axis, we obtain a *paraboloid*. This is widely used in radar to generate the narrow pencil beams required for tracking systems. Alternatively, we may use a portion of the paraboloid, making the horizontal aperture greater than the vertical aperture. The resulting fan beam is suitable for surveillance radars (see figures 1.2 and 3.7).

The beamshape can be modified by making small adjustments to the para-bolic contour. Cosecant-squared cover is often approximated in this way. The lower part of the parabola is bent slightly upwards, directing energy towards the higher elevation angles. This is shown by figure 4.5b. An alternative approach is to use two energy sources, or *feeds*. Careful positioning gives a composite beam close to the desired pattern.

Feeds positioned as in parts (a) or (b) of the figure pose two problems. Firstly, the transmission and feed system causes *aperture blocking*. Some of the reflected energy is intercepted and the beam pattern is degraded. Secondly, a portion of the reflected energy re-enters the feed as a *reverse wave*. The efficiency of the transmission system is affected by the resulting mismatch. Both problems are largely eliminated by using the *offset feed* arrangement of figure 4.5c. The feed,

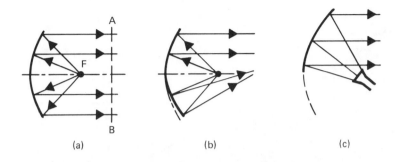

Figure 4.5 Parabolic reflectors

again placed at the focus, is tilted upwards and the lower portion of the reflector is removed.

Most microwave radars use hollow waveguides to connect the transmitter-receiver to the antenna. As far as the antenna feed is concerned, it is common to employ a *waveguide horn*. This is a carefully formed, flared opening at the end of the waveguide, with good directional and impedance-matching properties. The horn is often fitted with a circular polariser to reduce precipitation echoes. It may also incorporate a low-noise RF amplifier as the first stage in the receiver chain. It is good design practice to place the RF amplifier as near the antenna as possible, reducing the effects of noise and losses in the transmission system.

Not surprisingly, practical antenna feeds — including waveguide horns — never appear as true point sources. This fact, together with diffraction effects at the edge of the reflector, means that radar beams are not formed in quite the idealised way shown in figure 4.5. Nevertheless, the figure is useful in showing the basic principles underlying a parabolic antenna.

Reflector surfaces may be made from sheet metal, metal tubes or mesh. The latter are very economic, and offer low wind resistance. However, some energy tends to leak through the surface, increasing the antenna backlobe. Modern plastic materials and carbon fibre composites are increasingly specified for their lightness, rigidity, accuracy and resistance to corrosion (see figure 3.7).

Antennas which are subject to high winds, icing or extremes of temperature may be enclosed in *radomes*. A radome gives protection to the antenna itself, and eliminates wind load fluctuations on the drive motor and gearbox unit. Some form of radome is invariably needed in an airborne radar for aerodynamic reasons. The most common type of radome for ground-based installations is the *geodesic dome*. This is constructed from a load-bearing lattice, filled in with thin dielectric panels.

We should now consider the question of beam formation in rather more detail. The first point to note is that the radiation pattern of an antenna is not fully developed at short distances. In the *Fresnel region*, rays from the antenna

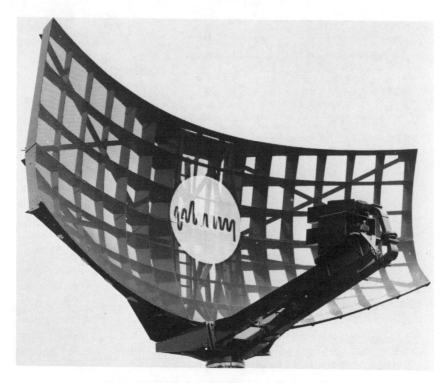

Figure 4.6 Antenna of the *Watchman* S-band radar, designed to produce
            cosecant-squared vertical coverage. The upper horn is used for trans-
            mission and reception, the lower horn for reception only. Circular
            polarisers and a low-noise FET RF amplifier are fitted. Horizontal
            beamwidth is about 1.5° (photo courtesy of Plessey Radar Ltd)

to a point target are not truly parallel, and the pattern varies with distance. It is
only in the far-field, or *Fraunhofer region*, that the 'true' radiation pattern is
observed. The boundary between the two regions is not precise, but may be
taken as about a^2/λ away from the antenna. Here a represents the width of the
aperture. Thus for a typical terminal area radar having $a = 5$ m and $\lambda = 10$ cm,
the boundary is at about 150 m. Radar antennas are invariably designed to
operate in the Fraunhofer region.

It is important to realise that the beamshape of any antenna in the Fraunhofer region is determined by its *aperture distribution*. In other words, the
distant electric field pattern depends on the current distribution across the
aperture, in phase as well as amplitude. To understand beam formation by a
parabolic reflector, we need to examine this idea rather more fully.

Consider the idealised rectangular aperture shown in figure 4.7. RF currents
are assumed to be flowing in the y direction, with intensity variations across the

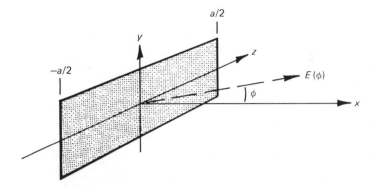

Figure 4.7 Beam formation by a rectangular aperture

aperture in the z direction. The variations are described by an aperture distribution function $A(z)$. The electric field intensity in the xz plane at some angle ϕ to the x axis may be found by summing contributions from current elements right across the aperture. Let us consider points in the Fraunhofer region for which $x \gg z$. Taking the aperture's centre ($z = 0$) as a reference, we see that as ϕ increases, the path length to current elements on one side of the aperture increases by $z \sin \phi$. The path length decreases by the same amount on the other side of the aperture. The corresponding phase changes are $\pm(2\pi z \sin \phi)/\lambda$ radians. Each contribution is weighted by $A(z)$ giving a far-field intensity of the form

$$E(\phi) = \int_{-a/2}^{a/2} A(z) \exp\left(j2\pi z \frac{\sin \phi}{\lambda}\right) dz \qquad (4.1)$$

Readers familiar with Fourier analysis will recall the *inverse Fourier transform* equation relating a signal waveform $f(t)$ to its spectrum $F(f)$

$$f(t) = \int_{-\infty}^{\infty} F(f) \exp(j2\pi ft) \; df \qquad (4.2)$$

These two equations are clearly similar in form. Indeed, the radiation pattern $E(\phi)$ and the aperture distribution $A(z)$ are related as a Fourier transform pair. This means that Fourier methods may be used to find the aperture distribution corresponding to a desired antenna beam pattern. The argument may be applied to elevation as well as azimuth by considering aperture distributions in both y and z directions.

Table 4.1 lists three forms of one-dimensional aperture distribution, and the main characteristics of their field patterns. Note that the beamwidth depends upon the ratio of wavelength to aperture (λ/a). A uniformly illuminated aperture gives the smallest beamwidth and largest gain, but it suffers from large sidelobes. (Incidentally, the radiated pattern in this case is of the $\sin x/x$ form familiar in signal theory.) A better compromise between beamwidth and sidelobe perform-ance is obtained with a *tapered distribution* such as the cosine or triangular function. Here the illumination decreases away from the centre of the aperture. All the distributions in the table are real functions of z, implying no phase vari-ations across the aperture. In the more general case, $A(z)$ would be a complex function of z.

Table 4.1

Form of distribution for $-\dfrac{a}{2} < z < \dfrac{a}{2}$	3 dB beamwidth (degrees)	Relative gain of main lobe (dB)	Sidelobe levels below main lobe (dB)	
			first	second
Uniform: $A(z) = 1$	$51\left(\dfrac{\lambda}{a}\right)$	0	-13.5	-18
Cosine: $A(z) = \cos\left(\dfrac{\pi z}{a}\right)$	$69\left(\dfrac{\lambda}{a}\right)$	-1.8	-23	-31
Triangular: $A(z) = 1 - \left\|\dfrac{2z}{a}\right\|$	$73\left(\dfrac{\lambda}{a}\right)$	-2.5	-27	-36

We should treat these theoretical values with some caution. They apply to simple analytical distributions for which the inverse Fourier transform is easily found. The design of practical antennas – including parabolic reflectors – is more likely to proceed in the opposite direction. That is, it starts with a desired specification for $E(\phi)$ and estimates the corresponding function $A(z)$. Computer-aided design is increasingly important in this field. It must also be remembered that manufacturing tolerances will tend to degrade the actual beam pattern, including the sidelobe levels. Nevertheless, table 4.1 is useful for its approximate values and for showing the effects of tapering. Practical parabolic reflectors with horn feeds commonly achieve first sidelobe levels of about -25 dB in the hori-zontal plane. Aperture distributions are often tapered down to about 10 per cent at the edges of the reflector. We should also note that the beamwidth is typically around 70 (λ/a) degrees. A surveillance radar with a horizontal beamwidth of $1.5°$ therefore requires an aperture about 50 wavelengths wide.

An important variation on the standard parabolic reflector is the *parabolic cylinder*, generated by moving a parabolic contour parallel to itself. The energy source must now be some form of *linear array*. It is placed at the focus of the parabola in the vertical plane, and runs nearly the full length of the reflector in the horizontal plane. The horizontal beamwidth is determined by the aperture illumination provided by the linear feed; the vertical beamwidth by the illumi-nation of the parabolic profile. The linear feed is often made from a waveguide with slots cut at intervals along its length. One advantage of the parabolic cylinder is that the radiation pattern can be controlled more or less independently in the two orthogonal planes. Careful design of the horizontal aperture illumination can yield very low sidelobe levels.

4.2.2 Phased arrays

As we have seen, the beam pattern of an antenna is determined by its aperture distribution $A(z)$. In the case of a parabolic reflector, $A(z)$ is continuous across the aperture. We have assumed in our previous discussion that beamshaping is obtained by suitable variations in the amplitude of $A(z)$, not the phase. An alternative approach is to build up an antenna from a number of individual radiating elements, giving a discrete aperture distribution. The elements may be any type of radiator – for example, dipoles or slotted waveguides. Phase variations between the elements are used to shape and control the beam, and to alter its direction in space. Such an antenna is called a *phased array*.

The phased array is inherently versatile. Phase control allows high-speed scanning without mechanical movement. Multiple beams may be formed with a single antenna, and continuously adapted to suit the operational situation. However, phased-array systems are very expensive. Their efficient use requires multichannel transmitting and/or receiving equipment, with real-time computer control.

The individual elements of a phased array are commonly arranged either in a straight line (a *linear array*) or in a plane (a *planar array*). Planar antennas give beam control in two dimensions. However, we may introduce the topic by discussing the simple linear array shown in figure 4.8. This has $(2m + 1)$ isotropic elements spaced d apart. We begin by assuming that there is no phase shift between the current in individual elements. However, when we consider the electric field intensity at a point in the Fraunhofer region, it is clear that contributions from the various elements are phase-shifted because of differences in path length. Denoting the shift between adjacent elements by θ_p, and the strength (amplitude) of the nth element by $A[n]$, we may write

$$E(\phi) = \sum_{n=-m}^{m} A[n] \exp(jn\theta_p) \tag{4.3}$$

where

$$\theta_p = 2\pi \left(\frac{d \sin \phi}{\lambda} \right) \tag{4.4}$$

Equation (4.3) is the discrete counterpart of equation (4.1) derived for the continuously illuminated aperture. Just as equation (4.1) has a similar form to the inverse continuous-time Fourier transform, so equation (4.3) has a similar form to the inverse discrete-time Fourier transform.

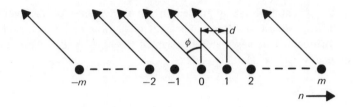

Figure 4.8   A linear array of isotropic radiators

In general, the individual radiating elements will have different strengths. However, for simplicity we here assume that $A[n] = 1$ for all n in the range $-m < n < m$. Equation (4.3) becomes

$$E(\phi) = \sum_{n=-m}^{m} \exp(jn\theta_p) = 1 + 2 \sum_{n=1}^{m} \cos n\theta_p \qquad (4.5)$$

which may be recast in the form

$$E(\phi) = \frac{\sin(\theta_p\{2m + 1\}/2)}{\sin\theta_p} = \frac{\sin(N\theta_p/2)}{\sin(\theta_p/2)} \qquad (4.6)$$

Here $N = (2m + 1)$ is the total number of array elements. $E(\phi)$ has a maximum value when $\theta_p = 0$, corresponding to $\phi = 0$. The main lobe of the radiated pattern is therefore normal to the array, in the *broadside* direction. As N becomes large, the pattern narrows and tends to the $\sin x/x$ form obtained with a uniformly illuminated rectangular aperture.

We next suppose that an additional phase shift θ_e is imposed between adjacent elements by electronic means. As far as the field pattern is concerned, the effects are indistinguishable from those due to path difference. Equation (4.6) therefore becomes

$$E(\phi) = \frac{\sin(N\{\theta_p + \theta_e\}/2)}{\sin\left(\dfrac{\theta_p + \theta_e}{2}\right)} \qquad (4.7)$$

The main lobe now occurs when $\theta_p + \theta_e = 0$, giving

$$\theta_e = -\theta_p = -2\pi \frac{d\sin\phi}{\lambda} \qquad (4.8)$$

or

$$\phi = \sin^{-1}\left(\frac{-\lambda\theta_e}{2\pi d}\right) \qquad (4.9)$$

These important results are illustrated in figure 4.9, for a linear array with 25 elements and a spacing $d = 0.5\lambda$. The line of the array is drawn vertically. We have

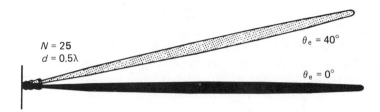

N = 25
d = 0.5λ
$\theta_e = 40°$
$\theta_e = 0°$

Figure 4.9 Beam-scanning by phase control

plotted $E^2(\phi)$ rather than $E(\phi)$ in order to show relative power gain, and have used linear scales. As expected, $\theta_e = 0$ gives a broadside beam. Choosing $\theta_e = -40°$ causes the beam to deflect about $13°$ from the normal, as predicted by equation (4.9). Variation of θ_e would clearly allow the beam to be scanned through a considerable angle.

The axial symmetry of figure 4.8 implies that a linear array of isotropic elements would produce its radiation pattern right around the array. A planar array with beam-forming in two dimensions must give a backward beam as well as a forward one. In practice, phased arrays use directive rather than isotropic elements, enhancing power radiation in the desired direction. Further beam-shaping and sidelobe reduction is normally achieved by tapering the strength of elements towards the edges of the aperture. Any backward lobe from a planar array may be effectively reduced by a reflecting screen.

Although we do not have space here for a detailed examination of equations (4.7) and (4.9), three general points should be mentioned. Firstly, it is found that if the element spacing d is too large, unwanted subsidiary beams are produced. These are called *grating lobes*. In practical systems d is usually made rather less than one wavelength, in order to avoid the problem. Secondly, the equations predict that beamwidth is a function of the scan angle. If the main lobe is scanned away from the broadwide direction, its beamwidth increases. The 3 dB beamwidth of a phased array with uniform aperture illumination is approximately

$$\theta_B \approx \frac{51\lambda}{Nd \cos\phi_0} \text{ degrees} \tag{4.10}$$

where ϕ_0 denotes the scan angle away from the normal. Fortunately the broadening effect is small for scan angles less than about $\pm 20°$. The final point concerns the idealised nature of our equations. We have assumed isotropic point sources, taking no account of mutual coupling, or diffraction and scattering of energy by adjacent elements. We cannot, therefore, assume that the equations will describe the performance of practical arrays with great accuracy.

Rotating planar arrays are quite often used for aircraft surveillance. Such systems allow discrimination in elevation as well as azimuth (and, of course, in range), and are often referred to as *3D radars*. The phased array is rotated mechanically to give coverage in azimuth. Pencil beam formation is achieved by amplitude and phase control of the array elements, with electronic scanning in elevation. The data rate may be improved by using multiple beams. It is often convenient to transmit a wide elevation beam which illuminates all targets on a particular azimuth, generating multiple beams on receive only. This may be done by suitable phase-shifting and summation of signals received by particular groups of array elements. The theme has a number of variations, serving to emphasise the versatility of phased arrays. We should, however, remember that operational versatility involves elaborate computer control and high capital cost.

Let us next consider how the variable phase shifts required by an electronically scanned phased array are generated. We note that the phase change produced by a transmission line of length l when an electromagnetic wave of frequency f_0 travels along it with velocity v is

$$\phi = 2\pi \frac{f_0 l}{v} \qquad (4.11)$$

The velocity depends on the permeability μ and permittivity ϵ of the transmission medium. We conclude that phase shifts may, in principle, be obtained

Figure 4.10 The 3D *Martello* phased-array radar, which operates at a wavelength of 23 cm. The planar array measures 7.1 m high by 12.2 m wide, and consists of 40 stacked horizontal linear arrays fed by a distributed solid-state transmitter. A single, cosecant-squared, elevation beam is transmitted. Eight elevation beams are synthesised on receive, using a passive beam-forming network. Elaborate signal processing facilities are incorporated (photo courtesy of Marconi Radar Systems Ltd)

by changing f_0, l, μ or ϵ. In practice, the first three methods have been widely used. We now make brief notes on each of these in turn.

Electronic scanning by frequency control is relatively simple to implement. Unfortunately, substantial frequency changes are needed to scan a beam over, say, $20°$. Much of the available bandwidth tends to be devoted to scanning, rather than frequency-agility or improvement of range resolution. The problem may be reduced in two ways. Firstly, we see from equation (4.11) that the *phase sensitivity* of the system to a change in frequency is given by

$$\frac{d\phi}{df_0} = \frac{2\pi l}{v} \tag{4.12}$$

This is proportional to l. Hence, by inserting an extra length of transmission line (normally waveguide) between each array element, and feeding the elements in series, it is possible to reduce the scanning bandwidth. Secondly, the *dispersive* property of waveguide transmission can be exploited. The propagation velocity through a waveguide is not strictly constant with frequency. In effect, this produces an additional, frequency-dependent phase shift which can be used to enhance $d\phi/df_0$. We should also note that frequency-scan systems sometimes use *within-pulse scanning*. The frequency is modulated within each transmitter pulse, rather than from pulse-to-pulse. Frequency diversity may also be employed, forming beams at a number of different centre frequencies. There is considerable flexibility in the design of frequency-scan radar.

The second main method of producing phase shift is to alter the line length l. The transmission line may be formed of waveguide, co-ax or stripline, built up from a number of cascaded sections. Individual sections are inserted or removed by electronic switching. Binary quantisation of line sections gives an economic method of building up a particular value of phase shift. The maximum quantisation error is, of course, determined by the total number of sections provided.

A third, and very popular, method is to use phase-shifters inserted in a waveguide. These two-port devices are commonly made of ferrite material. An applied DC magnetic field alters the ferrite's permeability, and hence the propagation velocity of microwaves passing through it. Phase shifters are not always bidirectional, or *reciprocal*. If non-reciprocal devices are used for both transmit and receive, the magnetic field must be rapidly switched between the two modes of operation.

It might be argued that only the last two methods – alteration of line length and the use of phase-shifters – give true phase control of a phased array. The frequency-scan method achieves its results indirectly. Sometimes a planar array uses phase-shifters to scan in one dimension, and frequency control to scan in the other. It may be referred to as a *phase-frequency system*, as opposed to a *phase-phase system* based entirely on phase-shifters or line-length control.

5 Signal Processing and Display

The processing and display of radar information have undergone major changes in the last 20 years, largely as a result of developments in microelectronics and computer technology. The detailed hardware of modern radar signal processors varies widely between different installations. However, certain key principles are now well established and widely implemented. We shall concentrate on explaining these principles in this chapter.

5.1 Matched filtering and pulse compression

Radar target echoes are generally contaminated by receiver noise or noise-like clutter. Before attempting to detect them using a threshold detector, it is important to improve the signal-to-noise ratio as much as possible. The optimum filter for achieving this is known as the *matched filter*.

A matched filter does not have to preserve the waveshape of the received echoes. The waveshape is known in advance, being more or less the same as that of the transmitted pulse. What is important is that the filter produces the largest possible peak output in response to each echo signal, for a given level of noise. In this way it maximises the chance of recognising valid echoes with a threshold detector.

The action of a matched filter is illustrated by figure 5.1. Part (a) shows two rectangular echo pulses. In part (b) they are contaminated by random noise. If this signal-plus-noise waveform is processed by a threshold detector, it is clear that occasional high noise peaks are likely to cause false alarms. Part (c) of the figure shows the effect of matched filtering. The peak output due to each signal is now considerably larger, compared with the noise, and a threshold level V_T can be successfully employed.

Note that the matched filter has not preserved the signal waveshape. (In fact, it gives a triangular form of output pulse if the input pulse is rectangular.) Furthermore, the peak signal output coincides with the trailing edge of the input pulse. Matched filter detection therefore involves a time delay equal to the duration of the input signal.

A matched filter is normally defined in terms of its impulse response, which always takes the form of a time-reversed version of the signal. Therefore such a filter is 'matched' to a particular signal waveshape – and to no other. It may be shown that, when this waveshape is fed into the filter, the output takes the form of the signal's *autocorrelation function*. For this reason a matched filter is often referred to as a type of *correlator*. The details of autocorrelation need not concern us here. It is enough to note that, whatever the form of the signal, the autocorrelation function is symmetrical with a peak value at its centre. It is this peak which we use for threshold detection.

Before describing such filtering in more detail, we must be clear that it cannot achieve the impossible. Some unwanted noise will inevitably appear at the filter output, and false alarms may still occur. All we can say is that a matched filter gives us the best chance of successful detection, for a given false-alarm probability.

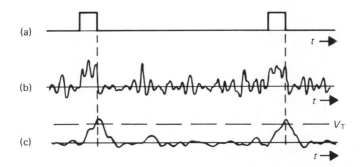

Figure 5.1 Improving signal-to-noise ratio by matched filtering

To understand the operation of a matched filter, we need to explore its action in the time domain. Figure 5.2a shows a time-limited signal 'pulse' $x(t)$, whose waveshape is assumed known in advance (we need not assume a rectangular pulse, because the concept of matched filtering is perfectly general and applies to any form of signal). The signal is passed into a matched filter, whose impulse response $h(t)$ is, by definition, a time-reversed version of $x(t)$. As the reader is probably aware, the impulse response represents the response to a unit impulse function $\delta(t)$ delivered at the instant $t = 0$.

As with any other linear filtering operation, the output from the matched filter may be found by *convolving* $x(t)$ with $h(t)$. Thus

$$y(t) = x(t) * h(t) = \int_{-\infty}^{\infty} x(\tau') h(t - \tau') d\tau' \tag{5.1}$$

where τ' is an auxiliary time variable. This convolution integral may be interpreted as a 'rolling together' of $x(t)$ and $h(t)$, after reversal of one of the functions. In the particular case of the matched filter, we have

$$h(t) = x(t_0 - t), \text{ giving } h(t - \tau') = x(\tau' + \{t_0 - t\}) \tag{5.2}$$

Hence

$$y(t) = \int_{-\infty}^{\infty} x(\tau')\,x(\tau' + \{t_0 - t\})\,\mathrm{d}\tau' \qquad (5.3)$$

The last equation shows that $y(t)$ is the integrated product of the input signal with a shifted version of itself. This is the signal processing operation known as autocorrelation, and $y(t)$ therefore has the same shape as the autocorrelation function of $x(t)$. Note that it has a peak value at its centre, and that the peak occurs after a delay equal to the duration of $x(t)$. The height of the peak may be found by substituting $t = t_0$ in equation (5.3), giving

$$y(t_0) = \int_{-\infty}^{\infty} x(\tau')\,x(\tau')\,\mathrm{d}\tau' = \int_{-\infty}^{\infty} x^2(\tau')\,\mathrm{d}\tau' \qquad (5.4)$$

This is simply a measure of the total energy in $x(t)$. Therefore the benefits of matched filtering depend only on the input signal's energy — not its detailed waveshape. It is now clear why our general form of the Radar Equation (see equation (2.24)) is expressed in terms of the average transmitter power P_{av}. When a matched filter is used in the receiver, target detectability depends on the average power, or energy, regardless of the transmitter waveform.

It may also be helpful to summarise the performance of the matched filter in the frequency domain. As with any linear time-invariant (LTI) system, its frequency and impulse responses form a Fourier Transform pair. In this case

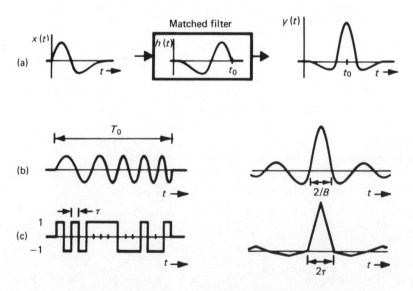

Figure 5.2 (a) The matched filter. (b) Pulse compression of a 'chirp' wave-form. (c) Pulse compression of a PRBS

$h(t)$ is a time-reversed version of the signal $x(t)$. Time reversal does not affect the magnitude of the corresponding spectral function, only the phase. So the frequency response magnitude must be identical to that of the input signal spectrum. Frequency components strongly represented in $x(t)$ are strongly transmitted; weaker components, less so. Noise lying outside the signal's band-width is completely rejected. In this way the matched filter optimises the signal-to-noise ratio.

As far as phase is concerned, the matched filter may be shown to have a phase response equal to the *complex conjugate* of the signal's phase spectrum. This has the effect of re-aligning the phases of the various components of $x(t)$ reaching the filter output. At a certain instant all the sinusoidal components of $y(t)$ reinforce one another in the time domain. Figure 5.2a shows that this occurs when $t = t_0$, giving rise to the required autocorrelation peak.

In a practical radar, matched filtering is accomplished at intermediate fre-quency. The filtering properties of the IF amplifier are therefore, ideally, those of the appropriate matched filter. For example, if a radar uses conventional 'rec-tangular' pulses, the impulse response of the IF amplifier should be a sinusoidal oscillation at intermediate frequency, of the same duration as an expected echo. In the frequency domain this corresponds to a narrow bandpass characteristic of $\sin x/x$ form. Although it is not possible to make a filter with precisely these characteristics, the use of an approximate filter leads to little loss of perform-ance. For example, a single-tuned *RLC* filter can give an improvement in signal-to-noise ratio only about 1 dB worse than the true matched filter. We therefore see that the filtering properties of the IF amplifier in a conventional pulse radar are not highly critical.

Matched filtering really comes into its own when pulse compression is used. We have already mentioned that pulse compression is increasingly specified for high-power radar systems. Transmitter pulses are stretched in time and specially coded. Received echoes are sharpened up, or compressed, by a matched filter prior to detection. The overall system therefore possesses the range resolution of a short pulse. A major advantage of the approach is that peak transmitter power may be greatly reduced, while maintaining average power. This is especially valuable in high-power, long-range radar applications. However, the matched filter must be realised with considerable accuracy.

What form of stretched pulse should be used? From a theoretical point of view many waveforms are possible. The main criterion is that the pulse should sharpen up well in the receiver — in other words, it should have a narrow, well-defined autocorrelation peak. In the frequency domain, this implies a substantial bandwidth. It is also desirable to spread the total energy evenly throughout the duration of the pulse. Therefore, the pulse envelope should not display large amplitude fluctuations. For this reason the waveforms used in practical compres-sion systems are normally either *frequency-coded*, or *phase-coded*.

A common way of implementing frequency-coding is to use *linear FM pulse compression*. The transmitter frequency is changed linearly with time over the duration T_0 of the pulse, as illustrated in figure 5.2b. (In practice, there would be hundreds, or thousands, of RF cycles within the pulse envelope.) In popular terminology, such a swept-frequency pulse is called a *chirp* waveform. Let us denote the frequency change through the pulse by B Hz. Then its ACF has a sin x/x form with a central peak of width $2/B$. Part of this function is shown on the right of the figure to a slightly expanded scale. If, for example, we require a compressed pulse duration of about 1 μs, we must expect to use a FM bandwidth of around 2 MHz.

Phase-coding may be achieved by dividing T_0 into a number of shorter intervals of length τ. The transmitter phase in each interval is set at either 0 or π radians. Once again the effect of coding is to increase transmitter bandwidth. A noise-like *pseudo-random binary sequence* (PRBS) is often used to determine the phase code. A so-called *maximal-length* PRBS, also known as an *m-sequence*, may readily be generated using a shift register with appropriate feedback connections. From a pulse compression viewpoint, m-sequences have well-defined autocorrelation properties with a large central peak.

An m-sequence with $N = 16$ binary characters is shown in figure 5.2c. We may think of its two levels (± 1) as determining the RF phase. The ACF, shown on the right, has a triangular peak and a number of smaller fluctuations to either side (only a few are shown here). Since the main peak width is 2τ, it is clear that a substantial compression effect requires subdivision of the transmitter pulse into many time-intervals. A code length between $N = 64$ and $N = 512$ would be more typical than the short sequence illustrated here. Recovery of the code in the receiver requires a phase-sensitive detector.

The *pulse compression ratio* (PCR) is a measure of the degree of compression achieved. It is the ratio between the duration of the transmitted pulse and the received pulse after matched filtering. It also indicates the increase in effective peak power of the transmitter, as a result of pulse compression. In linear FM pulse compression the PCR equals BT_0; with phase-coding, it is simply equal to N. Practical systems commonly achieve PCRs between about 50 and 500.

Although pulse compression offers excellent range resolution and reduced transmitter peak power requirements, it is not without its disadvantages. One of these is the occurrence of *time sidelobes* in the compressed pulse. The ACF waveforms in figure 5.2 show that such sidelobes occur to either side of the main peak. With linear FM coding, they follow a $\sin x/x$ function; with phase coding, they are triangular in form. In either case, there is some risk that they may be mistaken for separate target echoes, or that they will mask echoes from smaller targets nearby. Fortunately, a number of techniques have been devised for reducing time sidelobes, including weighting of the received pulses in either the time or frequency domain. Details are given in more advanced treatments of radar signal processing [see, for example, Skolnik (1980) – entry 1 in the Bibliography at the end of this book].

Another problem with pulse compression concerns minimum range. We have already noted that matched filter detection involves a delay equal to the duration of the signal – in this case, the stretched pulse. Furthermore, a pulse radar cannot receive echo signals until the transmitter pulse has ended. The system's minimum range is therefore limited by the transmitter pulse length. With a stretched pulse of, say, 50 μs duration, the minimum range is about 4 nm. This is unacceptable for many applications. A widely used solution is to transmit additional, short pulses for the detection of close-range targets. However, this increases the overall complexity of the system.

As we have seen, effective pulse compression requires transmission of carefully coded, stretched pulses. The usual approach is to generate the pulses at low power level and apply them, via suitable amplifier–mixer stages, to a RF power amplifier such as a travelling wave tube.

Let us now consider the practical problem of making a matched filter for pulse compression. We concentrate here on the commonly used chirp type of waveform. As already noted, matched filtering is normally carried out at intermediate frequency (IF) – typically between 30 and 70 MHz. Various filtering techniques are possible in principle; in practice, the *surface acoustic wave (SAW)* delay line has been widely adopted by radar designers. The basis of SAW device operation is illustrated by figure 5.3. Input and output transducers are mounted on a piezoelectric substrate such as quartz. An electrical signal, applied to the input transducer, generates an equivalent mechanical deformation which travels as an ultrasonic wave along the surface of the device. As the wave passes the *interdigital electrodes* of the output transducer, it produces an electrical output signal whose waveshape depends on the electrode geometry.

The operation of a SAW device as a matched filter may be visualised in two rather different ways. First, we may consider the chirp pulse in figure 5.2b as having a relatively low-frequency leading edge and a high-frequency trailing edge. If we pass it through a frequency-dependent, or *dispersive*, delay line, it is possible to achieve pulse compression by delaying the leading edge relative to the trailing edge. This is how the SAW device works. Its output transducer only provides efficient acoustic-to-electrical coupling if the electrodes are spaced apart by one half-wavelength of the propagating signal. The electrode spacing is continuously graded along the transducer's length and, in the example shown, lower-frequency components in the chirp waveform must therefore travel further along the surface before contributing to the output signal.

An alternative explanation is in terms of the device's impulse response $h(t)$.

Input transducer   Dispersive output transducer

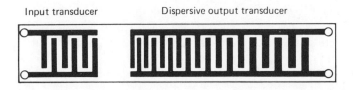

Figure 5.3 A surface acoustic wave (SAW) delay line

As previously explained, this must equal a time-reversed version of the signal to which it is matched. Therefore, in this example $h(t)$ must itself be a chirp waveform, but with a frequency which decreases towards the trailing edge. Let us now consider a brief impulsive signal launched by the input transducer. As the acoustic impulse passes each pair of output interdigital electrodes in turn, it generates an output fluctuation whose frequency is inversely related to the electrode spacing at that point. It is fairly easy to see that superposition of contributions from all the electrodes must give an impulse response of the required form.

Figure 5.3 is intended simply to illustrate the general principle. The reader is asked to bear in mind that various electrode geometries are possible. Furthermore, SAW filters may be used to compress pulses with other types of frequency or phase coding.

5.2 Clutter reduction

We have already commented several times on the problem of clutter, and have described its main technical characteristics in section 3.1. In most practical applications of radar, effective clutter reduction is a major preoccupation of the system designer.

Modern signal processing techniques can be highly effective against clutter. They may be split into two main categories: those which make use of the doppler shift produced by moving targets; and those which do not. In this section we concentrate on a number of techniques in the latter category. Being relatively simple to implement, they are often found in inexpensive radar systems.

First of all, we should be clear that clutter reduction is not exclusively the concern of signal processing. Prevention is often better than cure. In the case of ground-based or marine radar, several preventative measures have already been mentioned. They include

> The use of a small radar resolution cell
> Careful siting of the antenna
> Tilting the antenna beam
> The use of more than one antenna beam
>> (For example, two horn feeds are often provided with a parabolic reflector antenna, giving a high beam and a low beam. The former is less subject to clutter and is used to detect short-range or high-flying targets. The low beam is used for long-range targets which, because of the Earth's curvature, do not have to compete with significant ground or sea clutter.)
> The use of circular polarisation to reduce weather clutter

Moving on to signal processing, a very important technique for clutter reduction is swept gain, or sensitivity time control (STC). STC has been mentioned a number of times already (see particularly sections 2.1 and 3.1). It is often incorporated in the RF section of the receiver. The degree of swept gain may generally be selected to suit the clutter environment. In ground-based surveillance radars it is often made adaptive, and varies according to azimuth. Swept gain offers a very effective method of suppressing strong echoes from nearby clutter, although it does not improve the signal-to-clutter ratio.

Another important method for preventing clutter echoes from overwhelming an analog display is known as *log-FTC*. The receiver incorporates a logarithmic (or approximately logarithmic) input–output amplitude characteristic, followed by a *fast time constant (FTC)* circuit which acts as a high pass filter. The first of these features compresses the dynamic range of the detected clutter. The second removes its zero-frequency component, or mean level. The overall effect is to produce more or less constant clutter fluctuations at the receiver output, regardless of the strength of clutter at its input. (It may be helpful to mention that in Britain the term *short time constant* is often used in place of fast time constant, and abbreviated to STC. However, in this book we use FTC, reserving the abbreviation STC for sensitivity time control. This accords with American practice.)

The action of log-FTC may be explained as follows. Equation (3.10) gave the probability density function of the Rayleigh clutter envelope as

$$p(v_e) = \frac{v_e}{\psi_0^{1/2}} \exp\left(\frac{-v_e^2}{2\psi_0}\right), \quad v_e > 0 \tag{5.5}$$

This distribution has the property that its mean value $\psi_0^{1/2}$ and the amplitude of fluctuations about the mean are proportional to one another. The characteristic of a logarithmic receiver may be written in the form

$$v_2 = \alpha \log v_1 \tag{5.6}$$

where v and v_2 are its input and output voltage levels, and α is a constant. The gain G' of the characteristic to small fluctuations about a mean value $\bar{v}_1$ equals its slope at that point. Hence

$$G' = \left.\frac{dv_2}{dv_1}\right|_{\bar{v}_1} = \frac{\alpha}{\bar{v}_1} \tag{5.7}$$

Since G' is inversely proportional to the mean, and Rayleigh clutter gives an input fluctuation amplitude proportional to the mean, it follows that the output fluctuation amplitude must be constant. The output mean value may now be removed by a high pass filter. The resulting signal gives a roughly constant rate of false alarms (as a result of clutter) on a radar display, regardless of the input clutter level. The log-FTC technique therefore offers a simple form of *constant false-alarm rate (CFAR)* processing.

We must be careful not to overstate the relevance of equations (5.5) to (5.7). Practical clutter does not often have true Rayleigh statistics, nor can a practical receiver display a true logarithmic characteristic over its full dynamic range. Nevertheless, the log-FTC technique – or minor variations on it – has been widely used to reduce the effects of clutter on analog displays. Note that log-FTC, unlike STC, is independent of range.

In another important respect, log-FTC is similar to STC: it does not improve the target-to-clutter ratio. Neither of these techniques possesses what is termed *subclutter visibility*. In the case of log-FTC, constant false alarm rate is effectively traded against probability of detection of wanted targets. However, its CFAR capability can be valuable for preventing overload of the display by unwanted clutter. CFAR is particularly relevant to automatic detection and plot systems which tend to be overwhelmed by too many false alarms. We shall discuss more sophisticated versions of CFAR processing in section 5.6.

Instantaneous automatic gain control (IAGC) can also be used to reduce display overload caused by blocks of strong clutter. In many ways its action is similar to log-FTC. IAGC is implemented by controlling the gain of the IF amplifier with negative feedback. The time constant of the IAGC circuit is chosen so that short echoes from point targets suffer little or no attenuation; but echoes from extended clutter are reduced.

5.3 Digital signal processing

Perhaps the most striking recent development in radar technology has been the transition from analog to digital signal processing. This is particularly evident in high-performance systems incorporating such features as moving target indication and detection, automatic plotting and digital displays. However, as the years go by we may expect to see an increasing role for digital electronics in a wide variety of radar systems.

The reader will be aware that analog electronic circuits tend to suffer from the effects of temperature change and component ageing. Regular monitoring and adjustment is needed to maintain complex analog systems at peak performance. By contrast, digital circuits and computers, working on the binary principle, are remarkably stable and reliable. Another advantage of digital processing in the radar context is flexibility. Not only are elaborate signal processing schemes quite feasible, but they may be selected and modified under computer control. Radar systems can be made adaptive – for example, they can continuously adjust to variations in the clutter environment. The use of digital computers also allows information from a number of radars to be stored, processed and combined for presentation to a radar operator.

The conversion of radar signals to digital form is normally accomplished after IF amplification (including matched filtering/pulse compression) and detection. At this stage they are referred to as *video signals*, and have a typical bandwidth in the range 250 KHz to 5 MHz. The Sampling Theorem therefore indicates sampling rates between about 500 KHz and 10 MHz. Such rates are well within the capabilities of modern analog-to-digital converters (ADCs). Another way of considering sampling rate is in terms of range resolution. Suppose, for example, we require samples to be spaced the equivalent of 1/16 nm apart. Now 1 nm equals 1.85 km (see equation (1.2)). At a propagation velocity of 3×10^8 m s^{-1}, a go-and-return path of 1/16 nm corresponds to a time interval of 0.77 μs. Hence, in this case the required sampling rate would be about 1.3 MHz.

A certain amount of *quantisation noise* is always introduced by analog-to-digital conversion. Basically this is because a continuous-time signal is being represented by a finite set of discrete amplitude levels. Quantisation noise may be reduced by allocating more bits to the coding of each sample value. However, it is uneconomic to use a longer code than is warranted by the signal-to-noise ratio and dynamic range of the underlying signal. In many radar applications a code length of about 12 bits represents a good compromise. This gives 2^{12} = 4096 amplitude levels.

Analog-to-digital conversion of radar video signals is therefore quite straight-forward in principle. However, we should spare a few moments to consider the amount of high-speed storage required. A key application of digital processing is in moving target indication (MTI). Based on the doppler shift effect, it involves comparing successive echoes received from the same target. The radar signal must therefore be stored over at least one interpulse period. If we consider a medium-range system with a typical interpulse period of about 1 ms, and assume a sampling rate of 1.5 MHz, then it is clear that some 1500 sample values must be stored and continually updated. (We shall see later that this represents a fairly minimal requirement.) Bearing in mind that each sample may have a 12-bit code, we see that a large number of storage and delay elements are needed.

We now summarise a few key ideas about digital signal processing which will be useful in subsequent sections. In section 5.1 we noted that the impulse and frequency responses of an analog LTI system form a Fourier Transform pair. This is also true of a digital, or discrete-time, LTI system. The impulse response $h[n]$ is now defined as the system's response to the unit discrete impulse function $\delta[n]$. Figure 5.4 shows that $\delta[n]$ consists of a unit sample at $n = 0$, surrounded on both sides by zeros. In general, $h[n]$ consists of a series of *weighted* impulses, or samples, separated by the system sampling interval. Samples are usually binary-coded, and we may think of $h[n]$ as a sequence of numerical values.

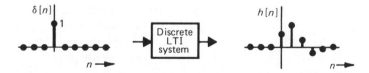

Figure 5.4 Impulse response of a discrete LTI system

The frequency response $H(\Omega)$ of such a system is given by the discrete-time Fourier Transform of $h[n]$. Thus

$$H(\Omega) = \sum_{n=-\infty}^{\infty} h[n] \exp(-j\Omega n) \tag{5.8}$$

If the system is causal, $h[n] = 0$ for $n < 0$, so the limits of summation become $n = 0$ to $n = \infty$. Equation (5.8) is the discrete-time counterpart of the continuous-time Fourier Transform used to describe analog signals and systems. The frequency variable Ω is equal to ωT, where ω is the usual angular frequency in radians per second and T is the sampling interval. Thus Ω is expressed in radians.

A very important feature of discrete LTI systems is that their frequency responses are always periodic in Ω, repeating at intervals of 2π along the frequency axis. This is an inevitable consequence of working with sampled signals. As we shall see in the following sections, it has major implications for radar signal processing.

5.4 Moving target indication (MTI)

5.4.1 The basis of MTI

The phenomenon used in MTI radar was mentioned in section 1.1 – targets moving with finite radial velocities produce doppler-shifted echoes. The doppler information can be used to discriminate in favour of moving targets such as aircraft, and against fixed targets and clutter. The principles of MTI were well understood by the mid 1950s, but it proved difficult to implement them with the available analog technology. Modern digital processing represents a major advance in this area of radar system design.

We start by considering the nature of the doppler information available to a pulse radar. Equation (1.1) gave the relationship between the doppler frequency f_d, target radial velocity v_r, and radar wavelength λ (or frequency f_0)

$$f_d = \frac{2v_r}{\lambda} = \frac{2v_r f_0}{c} \tag{5.9}$$

where c is the velocity of propagation (3×10^8 m s^{-1}). As we pointed out in section 1.1, doppler frequencies for aircraft targets typically fall in the range 0–5 kHz. However, it is important to remember that the doppler information is not continuously available to the signal processing system. It is only present during each brief received echo. In effect, the pulse radar 'samples' the doppler frequency of a particular target at the pulse repetition frequency f_p. This is illustrated by figure 5.5 in which, fairly typically, f_p is comparable with f_d. Note that the pulse length τ is extremely short compared with one period of the doppler. This means that we cannot expect to extract doppler information

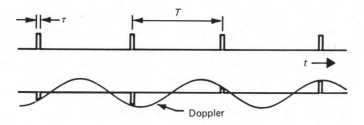

Figure 5.5 'Sampling' of target doppler frequency by a pulse radar

within any one received pulse. Instead, we must devise some means of comparing successive pulses.

The figure also helps explain the problem of blind speeds, first mentioned in chapter 1. If the doppler frequency produced by a moving target is exactly the same as the PRF (that is, $f_d = f_p$) then 'sampling' occurs at the same point on each doppler cycle. As far as the signal processor is concerned, it is as if the target were stationary. The same effect occurs if f_d is an integer multiple of f_p. Hence targets with certain radial velocities tend to be invisible to an MTI pulse radar. Referring back to equation (5.9), we see that these velocities must correspond to

$$f_d = \frac{2v_r}{\lambda} = nf_p \quad \text{giving } v_r = \frac{n\lambda}{2}f_p \tag{5.10}$$

where n is an integer. For example, consider a 10 cm terminal-area radar working at 1000 pulses per second. The blind speeds are given by

$$v_r = \frac{n \times 0.1}{2} \times 1000 = 50n \text{ m s}^{-1} = 97n \text{ knots} \tag{5.11}$$

Blind speeds at multiples of 97 knots would be operationally unacceptable. Fortunately there are ways of overcoming the problem, and we shall discuss them later in this section.

The blind-speed phenomenon may also be explained in terms of path length. When a target moves towards or away from the radar, the go-and-return path length alters from pulse to pulse, producing equivalent phase changes. However, if target range alters by an integral number of half-wavelengths ($n\lambda/2$) in one interpulse period ($T = 1/f_p$), the path length is changed by an integral number of *whole* wavelengths. Hence the phase of returning echoes, relative to the transmitter phase, remains fixed. The target appears to be stationary.

These comments on path length and phase change give important clues to the operation of practical MTI systems. Rather than attempt to measure the doppler frequency directly, we compare the phase of successive echoes from the same target using the transmitter phase as a reference. Phase changes from pulse to pulse denote a moving target; constant phase denotes a fixed target or clutter (or a target whose radial velocity is at a blind speed). Although the principle is quite straightforward, it is difficult to implement. The basic problem is that we require an extremely stable *phase reference*. For example, the 10 cm terminal area radar mentioned above has an interpulse period of 1 ms. The phase of each transmitter pulse must therefore be preserved, or 'remembered', for at least 1 ms, to allow comparison with the phase of returning echoes. This may not sound difficult – until we realise that a transmitter working at 3 GHz (λ = 10 cm) generates 3 million RF cycles per millisecond!

Figure 5.6 illustrates the classic *coho–stalo* method of extracting doppler phase shifts in a superheterodyne radar receiver. The coho (coherent oscillator) runs at the radar's intermediate frequency, denoted by f_c. It supplies the highly stable phase reference against which received echoes are compared in a *phase-sensitive detector (PSD)*. The stalo (stable local oscillator) runs at a much higher frequency f_1, such that $f_c + f_1 = f_0$, where f_0 is the transmitter frequency. It acts as a frequency mixer, converting up from IF to RF for the transmitter, and down again from RF to IF in the receiver. An important advantage of this arrangement is that any influences on transmitter phase due to the stalo are automatically compensated in the receiver. Although we have labelled the figure to show signal frequencies at various points in the system, we must remember that the doppler frequency f_d is not continuously available. For a particular target, it is only present during one pulse length τ in each interpulse period T. The reader will find many similarities between this figure and the more basic radar system shown in figure 1.1.

The PSD is essentially a mixer, supplied with IF by the coho, and doppler-shifted IF by the IF amplifier. It is unlike a normal envelope detector, since its output is proportional to the *phase difference* between the two inputs. We have already seen that this is one of the key operations to be performed in an MTI processor.

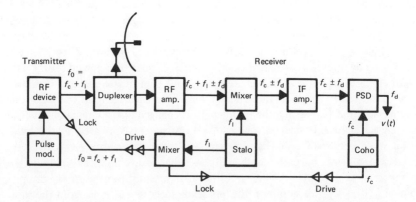

Figure 5.6   Extracting doppler information in a coho–stalo system

The precise action of the coho depends on whether the RF output device used in the transmitter is a power *amplifier* such as a klystron or travelling wave tube, or a power *oscillator* such as a magnetron. If an amplifier is used, the coho acts as a crystal-controlled master oscillator and phase reference which *drives* the output stage via the stalo mixer. This type of phase control is very stable and accurate and gives high-quality MTI. The system is said to be *coherent*. However, when a power oscillator is used, its phase cannot be controlled by an external drive signal, and tends to vary randomly from pulse to pulse. The coho phase must therefore be re-aligned, or *locked*, to that of the oscillator each time the transmitter fires. It is more difficult to achieve a highly stable phase reference in this way.

Figure 5.7 illustrates a typical output from the PSD, with a number of inter-pulse periods superimposed. The time origin corresponds to the instant when the transmitter fires. Receiver noise is not shown in the figure, for the sake of clarity. (Individual echoes appear as rounded, rather than rectangular, pulses. This is the typical effect of bandlimitation in the IF amplifier, which should ideally approximate a matched filter.) We also note that $v(t)$ can be either positive or negative, depending on the instantaneous phase difference between the incoming signal and the coho reference. $v(t)$ is therefore referred to as a *bipolar video signal*.

The figure clearly shows the difference between fixed and moving target echoes following phase-sensitive detection. A fixed target produces the same PSD output on every pulse. But a moving target gives an output which fluctuates at doppler frequency. The fluctuations produce a 'butterfly' effect when viewed on an oscilloscope. Such a video signal is not suitable for driving a radar display, because the fixed target echoes have not been eliminated. We now look at techniques for achieving this.

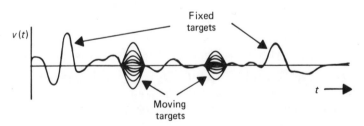

Figure 5.7   Fixed and moving target echoes after phase-sensitive detection

5.4.2 *Delay-line cancellers*

Since fixed-target echoes at the PSD output do not fluctuate in amplitude, we may use a subtraction process to discriminate against them. Thus, if $v(t)$ is subtracted from a version of itself which has been delayed by exactly one interpulse period T, fixed target echoes will *cancel*. But moving targets, which produce a different PSD output on each pulse, will remain.

A highly stable time-delay is difficult to achieve using analog techniques especially when we consider that T may be as much as a few milliseconds. Until about 1970, ultrasonic quartz delay lines were widely used. The modern approach is to sample the PSD output using an analog-to-digital converter, and to implement the delay and subtraction process digitally.

We must be careful not to confuse the two types of sampling discussed in this chapter. Firstly, there is the 'sampling' inherent in a pulse radar, illustrated by figure 5.5. Secondly, there is the much faster sampling used for digital signal processing, which must cater for the bandwidth of the radar video signal. Hence the delay T, when implemented digitally, must be built up from a large number of much shorter, individual delays – a point already discussed in section 5.3. Some form of electronic shift register is commonly used.

The block diagram of a digital subtractor is shown in figure 5.8a. The input signal $x[n]$ represents a sampled version of the PSD output $v(t)$. The output signal $y[n]$ equals the difference between two versions of $x[n]$, separated from one another in time by exactly one interpulse period T. In radar terminology such a system is known as a *single canceller*.

To understand the action of the single canceller as an MTI processor, we need to consider its impulse and frequency responses. If a unit impulse $\delta[n]$ is delivered on the input side, it is clear that the response must take the form shown on the right-hand side of figure 5.8a. Note that we are treating T as the effective sampling interval of the system. Thus

$$h_1[n] = \delta[n] - \delta[n-1] \tag{5.12}$$

Equation (5.8) may be used to give the corresponding frequency response

$$H_1(\Omega) = \sum_{n=-\infty}^{\infty} (\delta[n] - \delta[n-1]) \exp(-j\Omega n) \tag{5.13}$$

Using the sifting property of the unit impulse function, we may write directly

$$H_1(\Omega) = \exp(-j\Omega n)\big|_{n=0} - \exp(-j\Omega n)\big|_{n=1} = 1 - \exp(-j\Omega) \tag{5.14}$$

We are normally most interested in the magnitude of $H_1(\Omega)$, which is

$$\left| H_1(\Omega) \right| = \left| 1 - (\cos\Omega - j\sin\Omega) \right| = \{(1 - \cos\Omega)^2 + \sin^2\Omega\}^{\frac{1}{2}}$$

$$= \{2 - 2\cos\Omega\}^{\frac{1}{2}} = 2\sin\frac{\Omega}{2} = 2\sin\frac{\omega T}{2} \tag{5.15}$$

Therefore, the magnitude response of the single canceller is sinusoidal in form. It completely rejects zero-frequency signals, and also frequencies spaced by $2\pi/T$ radians per second, or $1/T$ Hz. Now $1/T$ Hz is just the radar's pulse repetition frequency f_p. We conclude that the single canceller, viewed as an LTI system or

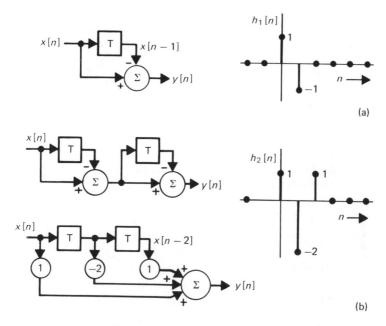

Figure 5.8 Delay-line cancellers

filter, rejects doppler frequencies equal to integer multiples of f_p. These correspond to the blind speeds of the radar (see equation (5.10)).

At the end of section 5.3 we noted that the frequency response of any discrete LTI system is repetitive in form. The single canceller may be considered
an example of this. It behaves as a sampled-data filter with a sampling interval
equal to T. Doppler frequencies corresponding to the blind speeds are rejected
because sampled sinusoids at these frequencies are identical to a sampled DC
level.

The frequency response magnitude characteristic of the single canceller is
drawn in figure 5.9. The horizontal axis of the figure is, in effect, a scale of
doppler frequency. Also shown are typical spectral characteristics of ground
and weather clutter. The ground clutter spectrum spreads around $f = 0$ because
of such effects as the swaying of trees and crops in the wind. The weather
spectrum shows not only a spread caused by the random motion of individual
cloud or rain droplets, but also a mean doppler shift caused by the prevailing
wind. We have previously discussed the spectra of moving clutter in section 3.1
and presented some typical examples in figure 3.2. Such spectra depend greatly
on the particular site and wind conditions. Note also that the clutter spectra
repeat along the frequency axis at intervals of $1/T$. Such *clutter foldover* is,
once again, an inevitable consequence of working with sampled data.

Figure 5.9 underlines the compromise inherent in the design of MTI cancellers. A narrow rejection notch at $f = 0$ (and at multiples of $1/T$) helps preserve

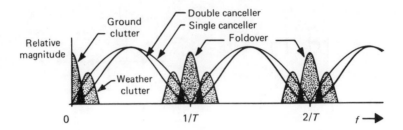

Figure 5.9   Frequency response magnitude characteristics of the single and
double canceller. Also shown are typical spectra of ground and
weather clutter

echoes from wanted targets which have small radial velocities (or velocities
close to the blind speeds); but a broad rejection notch is more effective against
moving clutter. No single design can be optimum for all radars in all weather
conditions. As far as the single canceller is concerned, we see that its rejection
notch is unlikely to be very effective against moving clutter. The *double canceller* characteristic, also illustrated in the figure, has often been preferred for
this reason.

A double canceller may be formed by cascading two single cancellers, as
shown at the top left of figure 5.8b. It is more complicated than the single canceller, requiring two delays equal to the interpulse period T. Its impulse response
$h_2[n]$, shown in the figure, is

$$h_2[n] = \delta[n] - 2\,\delta[n-1] + \delta[n-2] \tag{5.16}$$

The corresponding frequency response is therefore

$$H_2(\Omega) = \sum_{n=-\infty}^{\infty} (\delta[n] - 2\delta[n-1] + \delta[n-2])\exp(-j\Omega n)$$

$$= 1 - 2\exp(-j\Omega) + \exp(-2j\Omega)$$

$$= (1 - 2\cos\Omega + \cos 2\Omega) + j(2\sin\Omega - \sin 2\Omega) \tag{5.17}$$

The magnitude characteristic is given by

$$|H_2(\Omega)| = \{(1 - 2\cos\Omega + \cos 2\Omega)^2 + (2\sin\Omega - \sin 2\Omega)^2\}^{\frac{1}{2}}$$

which reduces to

$$|H_2(\Omega)| = 4\sin^2\frac{\Omega}{2} = 4\sin^2\frac{\omega T}{2} \tag{5.18}$$

$|H_2(\Omega)|$ is therefore the *square* of $|H_1(\Omega)|$ derived above. Its $\sin^2$ form gives
the broader rejection notch shown in figure 5.9. Actually, this result could have
been inferred directly. If we cascade two identical LTI systems, the overall
frequency response must be the square of that of each individual system.

An alternative way of implementing the double canceller is shown at the bottom left of figure 5.8b. It is fairly easy to see that such an arrangement of delay, multiplier and adder units produces the same impulse response $h_2[n]$. Realised in this form, the double canceller is a simple type of *tapped delay-line filter*, also called a *transversal filter*. The *tap weights* (in this case, 1, -2 and 1) are simply equal to successive impulse response values. Since the filter involves no feedback, it is said to be *nonrecursive*.

Confusion is sometimes caused by alternative names given to single and double cancellers. The single canceller, which has two terms in its impulse response, is also known as a *two-pulse canceller*; the double canceller, with three impulse response terms, as a *three-pulse canceller*. These names are commonly used when the canceller is arranged in the form of a transversal filter.

We have seen that the single canceller has a sin form of frequency response, the double canceller a $\sin^2$ form. It is quite possible to design a triple, or four-pulse, canceller with a response of $\sin^3$ form – and so on. The tap weights of higher-order cancellers may, in principle, be adjusted to give different compromises between clutter rejection and performance on moving targets. However, the modern approach towards enhanced doppler filtering tends to favour highly flexible frequency-domain methods. We shall cover these in section 5.5.

Before leaving our explanation of delay-line cancellers, two further points should be made. Firstly, such cancellers offer the advantage of processing all target and clutter echoes, regardless of range. There is no need for a separate filter for each range resolution cell. The complete received signal is processed, cell by cell, as it arrives.

The second point concerns the frequency responses shown in figure 5.9. These are theoretical steady-state responses, relevant to input signals which continue forever. However, in practice a pulse radar produces a limited number of hits-per-target on each scan of the antenna. Furthermore, the echoes vary in amplitude because the antenna gain falls off towards the beam edges. Both effects tend to degrade MTI performance, and mean that a delay-line canceller (or any other type of doppler processor) does not work under ideal steady-state conditions. We will return to this important matter in section 5.4.5.

5.4.3 Pulse repetition frequency (PRF) stagger

The next task is to consider how the problem of blind speeds can be overcome, or alleviated. Equation (5.10) shows that a blind speed occurs whenever target doppler-frequency is an integer multiple of the pulse repetition frequency (PRF). A moving target which is blind at one PRF will not, in general, be blind at another. The adverse effects of blind speeds may therefore be reduced by changing the PRF. This can be done from scan-to-scan, between groups of pulses, or from pulse-to-pulse. The latter technique is called *PRF stagger*.

A typical response of a single canceller system with PRF stagger incorporated is shown in figure 5.10. In this example the interpulse period is switched between three distinct values ('triple stagger') in the ratio 5:6:7. The radar wavelength is 23 cm and the mean PRF is 500 pps. Although scaled in radial velocity, the horizontal axis also denotes doppler frequency. The composite response is the average of the responses for each PRF considered separately. Hence, we are assuming that video integration (see section 2.2.2) causes an averaging effect as the antenna scans past an individual target. Note that the first blind speed now occurs when the blind speeds of the individual PRFs coincide – in this case at about 670 knots. This is much higher than it would be with a single PRF. Another important effect of PRF stagger is to make the low-frequency notch much narrower than with cascaded single cancellers having the same first blind speed. Whether or not this is an advantage depends on the severity of moving clutter.

The illustration is based on stagger ratios of 5:6:7. Closer ratios (for example, 11:12:13) could be used to give a higher first blind speed. However, the fluctuations in the response characteristic at intermediate radial velocities would then be more pronounced. The choice of stagger ratios is therefore a compromise. Double-stagger or multiple-stagger may also be specified, rather than the triple-stagger described here. In systems subject to jamming, random stagger may be used.

Although PRF stagger is a valuable technique, it has its disadvantages. Firstly, it increases system complexity. Extra delays must be switched in and out of the canceller, to compensate for changes in interpulse period. This was a serious matter in the days of quartz delay lines, although modern digital processing makes it far easier to achieve. Another disadvantage is that second-time-around clutter echoes cannot be cancelled, because they appear at different ranges from pulse to pulse. Where long-range clutter is a problem, it may be preferable to change the PRF from scan to scan, or each time the antenna scans through half a beamwidth.

When all has been said and done about blind speeds and PRF stagger, there is still the problem of *tangential fading*. An aircraft or other target moving tangentially to the radar produces no doppler shift. It appears stationary to the MTI processor, and its echoes are cancelled. Operationally this can be serious. For

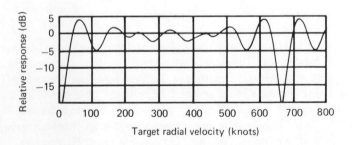

Figure 5.10 A PRF stagger characteristic

example, aircraft may be required to follow curved flight paths within a terminal area, and may fly tangentially part of the time. Note, however, that total loss of signal requires not only zero radial velocity, but also a constant target attitude with respect to the radar. Fortunately, this rarely occurs.

5.4.4 Digital MTI

Our description of MTI has so far made only general references to digital signal processing. We have noted that MTI cancellers nowadays use digital delay lines. And in the previous section we commented that PRF stagger is much simpler to implement digitally. However, digital techniques are not merely a replacement for analog ones; they greatly extend the power and flexibility of radar signal processing. A good example is the relative ease with which a dual-channel digital MTI processor tackles the problem of *blind phases*.

The blind-speed condition was summarised by figure 5.5, and for convenience we repeat the diagram in discrete-time form in figure 5.11a. The sinusoid represents the underlying doppler signal, and a blind speed occurs because 'sampling' by the pulse radar takes place at the same point in each cycle. (In fact, the *first* blind speed is illustrated here.) The rather different *blind-phase condition* is illustrated in part (b) of the figure. At the top, we see that the doppler signal is now sampled four times per cycle, producing the sequence $x_i[n]$. Suppose that $x_i[n]$ forms the input to a single canceller. Since alternate pairs of values are equal, the canceller's output will be zero on alternate pulses, representing an unwelcome loss of signal.

The blind phase effect is even more dramatic if sampling happens to occur twice per cycle, and coincides with the doppler zero crossings. In this case there is no output at all from the canceller. However, the figure is useful in suggesting the partial loss of signal which occurs more commonly in practice.

Next suppose that the same sampling instants are used on a second version of the doppler signal, phase-shifted by $90°$ with respect to the first. The resulting sequence $x_q[n]$ is also shown in the figure. Once again, alternate pairs of values are equal; but they are not the *same* pairs as in $x_i[n]$. Applied to its own single

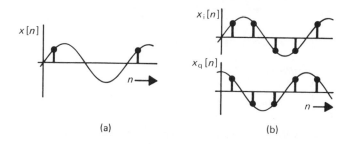

(a) (b)

Figure 5.11 Blind-speed and blind-phase conditions

canceller, $x_q[n]$ will therefore produce a finite output whenever $x_i[n]$ produces zero output, and *vice versa*.

Figure 5.11b uses a particular set of phase and frequency relationships to illustrate blind phases. In practice, the effect is rarely as clear as this. Nevertheless, there is still an advantage to be gained by using twin MTI channels.

A block diagram for such a system is given in figure 5.12. The IF amplifier now feeds two phase-sensitive detectors. The coho signal to one of them is phase-shifted by $90°$. The in-phase (I) and quadrature (Q) channels are otherwise identical, each containing an analog-to-digital converter and a single canceller, implemented digitally. The canceller outputs are combined prior to digital-to-analog conversion (alternatively, further digital processing may be performed). The combination law indicated, $(I^2 + Q^2)^{\frac{1}{2}}$, is the most effective. However, in practice the simpler alternative $(|I| + |Q|)$ has also been quite widely used.

In principle, it is possible to implement dual-channel MTI using analog techniques. However, it has rarely been attempted in practice, because of the instabilities associated with analog delay lines and circuits. By contrast, the stability, reliability and decreasing cost of digital processing make the provision of the extra channel an attractive option.

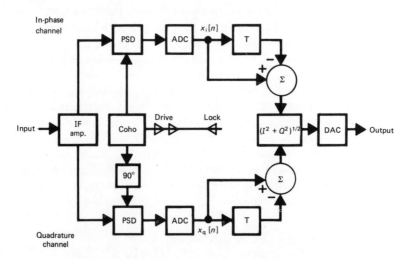

Figure 5.12   Dual-channel digital MTI

5.4.5 *The MTI improvement factor*

We next consider the factors affecting overall performance of a practical MTI system. Unfortunately, this is a highly complicated question. We have seen that MTI processing involves various design options – different types of cancellation, PRF stagger, single-channel or dual-channel operation, and so on. Since no two MTI radars are likely to share the same basic operating parameters, let alone the same clutter environment, it is very difficult to make meaningful comparisons between them. Nevertheless, some overall assessment of MTI performance is commonly required, and attempted.

The most widely used measure is the *MTI improvement factor*, defined as

$$I = \frac{(S/C)_{out}}{(S/C)_{in}} \tag{5.19}$$

where $(S/C)_{in}$ and $(S/C)_{out}$ denote the signal-to-clutter ratios at input and output of the MTI processor respectively. They are averaged over the complete range of doppler frequencies, and hence target radial velocities. (We must not confuse the symbol I with its use to denote in-phase MTI channel, as in the previous section.) The improvement factor is usually expressed in decibels. Since it represents the ability of the processor to reject clutter in favour of moving targets, its value must obviously depend on the processor's frequency response characteristic. It is also affected by other factors, some of them statistical. They include:

> The finite number of hits-per-target
> The clutter environment, including clutter motion
> Equipment instabilities
> The code length used for digital signal representation

We now make brief notes on each of these in turn.

Earlier in this section we noted that the canceller frequency responses in figure 5.9 apply to ideal steady-state conditions, whereas a radar with a rotating antenna produces a finite number of hits-per-target. The MTI processor must therefore work on a limited set of echoes from any one target. A second, closely related effect is the amplitude modulation of successive echoes caused by antenna rotation. The largest echo is received when the beam axis points directly at the target. Smaller echoes are received from the beam edges. The effect is called *scanning modulation*. It degrades cancellation performance on fixed targets or clutter, giving an upper limit to the MTI improvement factor which can be achieved. Some approximate figures are given in table 5.1.

Table 5.1

Hits-per-target (n)	Maximum MTI improvement factor (dB) using	
	single cancellation	double cancellation
10	18	34
30	28	53
100	38	73

As far as the clutter environment is concerned, we have already dealt, at some length, with the basic characteristics of clutter in section 3.1. Figure 3.2 showed typical doppler spectra for moving clutter relevant to a radar frequency of 1 GHz, and in equations (3.13) and (3.14) we defined the rms velocity spread σ_v. The general problem posed by moving clutter to an MTI canceller was illustrated in figure 5.9. Using such information it is possible to predict the maximum possible MTI improvement factor in the presence of moving clutter [see, for example, Skolnik (1980) – entry 1 in the Bibliography at the end of this book]. The appropriate expressions may be written as

$$I_1 = \frac{\lambda^2}{8\pi^2 \sigma_v^2 T^2} \text{ for a single canceller} \tag{5.20}$$

and

$$I_2 = \frac{1}{2} \left(\frac{\lambda^2}{8\pi^2 \sigma_v^2 T^2} \right)^2 \text{ for a double canceller} \tag{5.21}$$

Table 5.2 lists values of these expressions for five types of moving clutter shown in figure 3.2, estimated for a 10 cm radar working at a PRF of 1000 pps ($T = 1$ ms).

Table 5.2

Type of clutter	Maximum MTI improvement factor (dB) using	
	single cancellation	*double cancellation*
(a)	55	107
(b)	35	67
(c)	22	41
(d)	21	39
(e)	15	28

The influence of radar wavelength and interpulse period on equations (5.20) and (5.21) shows that similar values should apply to a 23 cm long-range surveillance radar working at about 400 pps.

It must be stressed that the values in table 5.2 are very approximate. They assume that the mean clutter velocity (and hence doppler shift) is zero. And they apply to a uniform clutter environment, whereas radar clutter is usually variable in range and azimuth. For these reasons, the table should only be used as a rough guide to the effects of clutter motion on MTI performance.

The third factor we should consider is equipment instability. Almost any instability in the transmitter/receiver system may cause the apparent frequency spectrum of stationary clutter to broaden, leading to a degradation of cancellation performance. We have already mentioned the importance of highly accurate and stable time-delays in a canceller, and the benefits of a driven power amplifier in the transmitter for achieving an accurate pulse-to-pulse phase reference. Fortunately, the trend towards digital signal processing, and digital timing and control of radar transmitters, offers major advances in stability compared with the former analog techniques. We cannot discuss this matter fully here. However, it is worth noting that the limitation on MTI improvement factor resulting from equipment instability is commonly quoted in the range 50–70 dB for modern ATC surveillance radars. (Most such radars use MTD-style processing as described in the following section, rather than delay-line cancellers.)

Related to the topic of equipment instability is the binary code length used in a digital processor. We have mentioned the quantisation noise produced by analog-to-digital conversion in section 5.3. Once introduced, such noise cannot be removed. It represents an unwanted 'instability' in the amplitude coding of individual signal samples. Theoretically, it limits the maximum MTI improvement factor to slightly less than $6N$ dB, where N is the code length in bits. For example 9-bit coding gives about 52 dB; the widely used 12-bit coding, about 70 dB.

How can all these factors be combined to produce a realistic value of I for a practical radar? A very important point is that I is limited by the worst of the contributing factors. It is rather like building up an audio system from amplifiers, loudspeakers, and so on. There is little virtue in spending a lot of money on one part of the system if performance is severely limited by another part – or by extraneous acoustic noise. Similarly, the radar designer must balance the difficulty and cost of achieving a given level of equipment stability against limitations imposed by scanning modulation and clutter movement. It is hardly an exaggeration to say that this requires the judgement of a Solomon.

An alternative measure of overall MTI performance, used less today than formerly, is *subclutter visibility (SCV)*. SCV describes the ability of a radar to detect a weak target echo in co-incident clutter. All target radial velocities are assumed equally probable. Thus an SCV of 30 dB means that a moving target is detectable in clutter even though its echo power is 1000 times weaker than that of the clutter. As we noted in section 5.2, techniques such as STC and log-FTC do not possess SCV because they do not improve the signal-to-clutter (S/C) ratio.

The distinction between SCV and the MTI improvement factor I becomes clear if we note that I is solely a measure of the MTI system's ability to *improve* the S/C ratio. It does not tell us whether the ratio, after MTI processing, is adequate to achieve stated detection and false-alarm probabilities. SCV, on the other hand, is quoted for particular values of these probabilities, and includes the final detection of wanted targets. If, for example, $I = 45$ dB and we require an S/C ratio of 12 dB after MTI processing, then the SCV is $(45 - 12) = 33$ dB.

Although this section has concentrated on the performance of single and double cancellers, much of the discussion is relevant to other doppler-processing techniques. These include the MTD style of processing described in the following pages.

5.5 Moving target detection (MTD)

5.5.1 The basis of MTD

During the early 1970s a new type of radar signal processor was developed at the Massachusetts Institute of Technology for airport surveillance radars in the United States. It was called the Moving Target Detector. There are two main differences between MTD-style processing and the more traditional MTI techniques described in the previous sections. Firstly, MTD makes extensive use of digital signal processing. This allows it to adapt continuously to the clutter environment, enhancing the visibility of moving targets at the expense of fixed and moving clutter.

The second main difference is in the way that doppler filtering is carried out. An MTD system performs a frequency analysis of incoming signals, using the technique of *discrete Fourier Transformation (DFT)*. This is most efficiently implemented using a *fast Fourier Transform (FFT)* algorithm. In effect FFT analysis resolves the doppler signal into a number of separate spectral bands, and offers much greater processing flexibility.

The DFT of a discrete-time signal $x[n]$ having N sample values is defined as

$$X_k = \sum_{n=0}^{N-1} x[n] \exp(-j2\pi kn/N) \tag{5.22}$$

with the integer k taken from 0 to $(N - 1)$. In effect, the signal is analysed into N elementary spectral bands. If the value of k is taken outside these limits, it is found that the coefficients X_k form a periodic sequence. This emphasises the fact that the frequency-domain description of any sampled-data signal or system repeats indefinitely along the frequency axis.

As it stands, equation (5.22) involves a considerable amount of redundant computation. The exponential term is a periodic function with a limited set of values for the various combinations of n and k. So if the DFT is implemented directly, the same values tend to be calculated over and over again. FFT algorithms aim to eliminate this redundancy, and give considerable increases in speed. They are most efficient when the number of samples to be transformed (N) is an integer power of 2. Their speed advantage over the standard DFT, based on the number of multiplications involved, is of the order $N/\log_2 N$. FFTs may, in principle, be implemented either in software or in hardware. However, the speed requirements of radar signal processing demand dedicated FFT hardware, and special-purpose integrated circuits are available.

We should be careful not to imply that FFT processing of a radar signal is somehow essentially different from filtering based on delay-line cancellers. FFT analysis is, in fact, equivalent to passing the signal into a bank of elementary bandpass filters. For this reason, a radar FFT processor is often referred to as a *doppler filter bank*. A similar result could certainly be obtained using a set of digital bandpass filters based on delay, multiplier and summing units. However, it would be far less economic.

The frequency response characteristics of a typical doppler filter bank are illustrated in figure 5.13, for the case $N = 8$. Each filter response repeats indefinitely along the frequency axis at intervals equal to the radar's PRF ($1/T$ Hz). Individual responses are of sin x/x form – although we have omitted sidelobes from the figure, for the sake of clarity. There is considerable overlap between adjacent responses, which can hardly be described as 'ideal' bandpass functions. Nevertheless, the filter bank provides a basic frequency analysis of an incoming signal. If it is required to reduce the sidelobe levels of the individual filters (at the expense of some increase in main lobe width), this can be achieved either by tapering, or *windowing*, the signal samples in the time domain – or by equivalent spectral weighting in the frequency domain. Such techniques are analogous to tapering the aperture illumination of a radar antenna in order to improve its sidelobe performance (see section 4.2.1).

The main practical advantages of FFT processing of radar signals may be summarised as follows

(1) Division of the doppler frequency domain into N separate bands offers a very flexible approach towards discriminating against fixed and moving clutter. If moving clutter (such as that from weather or birds) appears with a non-zero mean doppler shift, the thresholds at the outputs of the various doppler filters may be raised accordingly. The system can therefore be made *adaptive* to the clutter environment, rejecting clutter which would be passed by the usual type of delay-line canceller.

(2) Although the filter centred at zero frequency (and at the PRF and its harmonics) has no clutter-rejection capability, it may be used to generate

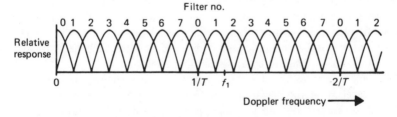

Figure 5.13 Frequency responses of a doppler filter bank

a *clutter map*. This offers a valuable method for detecting targets moving tangentially to the radar, as described later.

One difficulty posed by the repetitive nature of each doppler filter's response is that a fast-moving target may occupy the same doppler filter as fixed or moving clutter, and be masked by it. For example, an aircraft producing the doppler frequency f_1 in figure 5.13 appears strongly at the output of filter no. 1. Since the same filter transmits doppler frequencies close to zero, it may simultaneously have a large output due to weather clutter. Fortunately, the problem may be overcome by alternating between two (or more) different values of PRF. A wanted target will, in general, appear at different filter outputs in the two cases, and be visible in at least one of them. The technique also allows the radial velocity of targets to be unambiguously determined. It has clear parallels with PRF stagger, as used with delay-line cancellers. However, it is impracticable to change the PRF on a pulse-to-pulse basis when using FFT processing, so it is commonly altered every half-beamwidth, or from scan to scan.

We should consider how many individual filters are required in a doppler filter bank. It might seem desirable to divide the frequency scale up very finely, giving a high degree of discrimination between targets and clutter with different doppler shifts. But we must remember that a doppler filter, like a delay-line canceller, has to work with a limited number of hits-per-target. Decisions about the presence or absence of a target (or clutter) may typically have to be made on the basis of between 10 and 20 returning echoes. For this reason, radar FFT processors are often designed to work on groups of either 8 or 16 signal samples (remember that the FFT is most efficient when N is an integer power of 2). This, in turn, produces doppler filter banks having either 8 or 16 individual filters.

5.5.2 Adaptive MTD systems

Practical moving target detectors involve much more than a doppler filter bank. Although there is some variation between installations, and between different radar manufacturers, the main components of a typical system are as shown in figure 5.14.

We first note that the output of filter 0 in the doppler filter bank tends to be dominated by fixed clutter echoes having zero doppler shift. By not using this output, fixed clutter may be suppressed. The suppression is not complete, however, because adjacent filters have significant sidelobes in the region of zero doppler frequency. To reduce such clutter 'crosstalk', an MTI canceller is included at the input side of the FFT processor. This has the additional benefit of reducing the dynamic range of signals entering the filter bank.

FFT analysis is followed by frequency-domain weighting to reduce filter sidelobe levels, and the magnitude of the output in each spectral band is computed. A separate threshold level is applied to each filter output, to test for target echoes. The threshold is determined by the system noise level, and by any

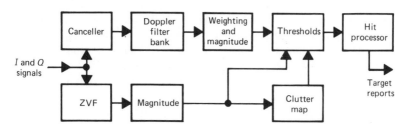

Figure 5.14 Block diagram of an adaptive MTD processor

echoes appearing at the output of the same filter in a number of adjacent resolution cells. Thus, the threshold takes account of the average level of moving clutter (such as rain) in the vicinity of the cell under test. The use of such an *adaptive video threshold* for each doppler filter in each resolution cell provides constant-false-alarm-rate processing. We discuss this more fully in the next section.

A zero-velocity filter (ZVF) is also shown in the block diagram. Its function is to recover clutter echoes suppressed by the MTI canceller, using them to generate a *clutter map*. The map is built up over a number of antenna scans (typically between 6 and 20), and is continuously updated. It therefore adapts to slow changes in weather or ground clutter. The values stored in the map are used to establish thresholds for targets with zero radial velocity. In this way the problem of tangential fading suffered by traditional MTI cancellers is largely overcome. The technique is said to provide *super-clutter visibility*.

As the antenna scans past any one target, there may be threshold crossings at one or more doppler filter outputs, in several or many interpulse periods, and perhaps in adjacent resolution cells. The function of the *hit processor* is to correlate all threshold crossings, grouping together those which appear to come from the same target. At this stage, signal amplitude and doppler shift information is used to eliminate very weak echoes, or those judged to come from moving clutter or angels. The hit processor generates target reports comprising range, azimuth, amplitude and radial velocity information, on all validated targets. It may also produce a reconstituted (synthetic) video signal for presentation on radar displays.

The block diagram of figure 5.14 does little to suggest the enormous amount of digital storage and signal processing involved in such a scheme. A typical system may divide each interpulse period into a thousand or more range intervals, and use several hundred azimuth intervals. Data from between 10 and 20 successive interpulse periods must be stored, and presented, one resolution cell at a time, to the doppler filter bank. The overall radar output is therefore divided into millions of individual range–azimuth–doppler cells, each of which has its own adaptive threshold. The clutter map may have 100 000 or more range-azimuth cells, and is continuously updated. And we have not even mentioned the canceller, zero-velocity filter and hit processor shown in the figure! It hardly needs saying that this extraordinary processing sequence depends crucially on recent advances in microelectronics and computing. The reader who is interested in the practical application of MTD will find further information in sections 7.2 and 7.5.

Although full digital MTD processing is technically feasible, it is also expensive. Various simpler schemes, incorporating a clutter map and aspects of MTI/ MTD filtering, may be used instead. Typically, three parallel channels of signal processing are provided, and a target declared following validation on one or more channels. The channels are:

> *Normal radar channel*, without MTI/MTD, for detecting targets 'in the clear'
> *Adaptive ground-clutter filter* for detecting targets in the presence of stationary clutter
> *Adaptive moving-clutter ('auto-doppler') filter*, for detecting targets in the presence of moving clutter — principally weather

In section 5.4.5 we discussed the improvement factor I, used to quantify the enhancement of signal-to-clutter ratio as a result of MTI processing. The same performance index is widely used for MTD. Once again, it is an average measure taken over the full range of doppler frequencies, and is limited by the number of hits-per-target and equipment instability. Values in the range 40–60 dB are typically claimed for ATC radars detecting moving targets in fixed clutter, reducing to 25–35 dB for targets in moving clutter.

5.6 Constant false-alarm rate (CFAR) processing

CFAR processing aims to produce a constant and acceptable mean rate of false alarms, regardless of the levels of noise and clutter. We have already made two references to CFAR in this chapter. In section 5.2 we saw that a log-FTC characteristic offers a simple form of analog CFAR processing when the noise or clutter, after envelope detection, has Rayleigh statistics. And in section 5.5.2 we introduced the idea of an adaptive video threshold, noting that it could be used to provide CFAR in a digital MTD system.

Some form of CFAR is normally essential when radar signals are processed digitally, and presented on digital displays. However, in simpler types of radar it may be unnecessary. When a radar signal is viewed on a conventional analog display, false alarms caused by noise or clutter appear directly as unwanted 'blips' on the screen. The detection threshold is effectively the signal level which just causes the screen phosphor to brighten. An experienced operator becomes skilful at recognising true targets, and if he is confused by too many false alarms the gain control on the display can be turned down. In a sense, the operator provides his own CFAR processing – although it cannot, of course, compensate for rapid clutter fluctuations.

The modern trend is towards digital techniques which provide automatic detection and tracking (ADT). Decisions about the presence of targets are removed from the human operator. Radar personnel are presented with 'clean' displays, driven by synthetic computer-generated signals. In such cases CFAR is essential, because an ADT computer can easily become overwhelmed by too many false alarms. The detection threshold must adapt automatically to both short-term and long-term changes in clutter and noise levels. The critical nature of the problem becomes clear if we refer back to figure 2.7. In a typical radar receiver a change of only 1 dB in the threshold-to-noise (or clutter) ratio can alter the false alarm rate by about 2 orders of magnitude.

CFAR processing may be desirable, or essential, in many radar applications. But it has its disadvantages. The main one is that control of false-alarm rate in regions of dense clutter inevitably causes some loss of wanted targets. This is a reflection of the compromise between false-alarm probability and target-detection probability, discussed in some detail in section 2.2.2. In addition, if a radar is subject to jamming or interference from other radars, its CFAR circuits may raise the detection threshold even higher – without the operator necessarily being aware of it. In such cases it is important to use special anti-interference measures before subjecting the signal to CFAR processing.

We must also remember that CFAR techniques do not have subclutter visibility. They make no special distinction between threshold crossings caused by noise, clutter, interference or wanted targets. Their aim is simply to restrict the average rate of threshold crossings to a manageable value.

Various CFAR processing techniques have been devised. We have already mentioned log-FTC, as used with analog systems and displays. However, one of the most important and versatile techniques is *cell-averaging CFAR*, which provides an adaptive video threshold. It is widely employed in digital processing schemes including MTD.

A typical cell-averaging CFAR processor is illustrated in figure 5.16. Tapped digital delay lines are used to estimate the average level of noise or clutter to either side of the range cell of interest, or *test cell*. The various tap outputs are summed, and the result used to set the detection threshold for the test cell. If there are high levels of noise or clutter in the surrounding cells, the threshold is set high – and *vice versa*. The threshold law must be carefully devised to give an acceptably low average false-alarm rate, under variable clutter conditions.

In the scheme illustrated, the range cells ahead of the test cell are summed separately from those behind it. This allows the threshold to be determined by the *greater* of the two sums, reducing generation of false alarms at the leading and trailing edges of sharply defined clutter regions.

Several variations on the cell-averaging theme are possible. More taps may be used, giving improved estimates of average noise and clutter, and making the

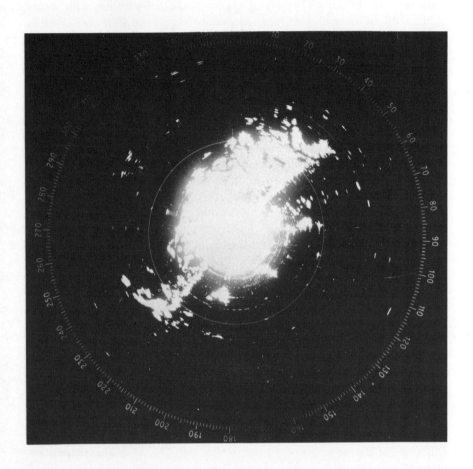

(a)

Figure 5.15   Typical PPI displays of a 10 cm ATC radar, with and without
              signal processing. (a) A normal ('raw') radar picture. (b) A processed
              display showing about 50 aircraft tracks. Note also the synthetic
              map data and range ring. The processing includes adaptive ground
              and moving-clutter filters (photos courtesy of Plessey Radar Ltd)

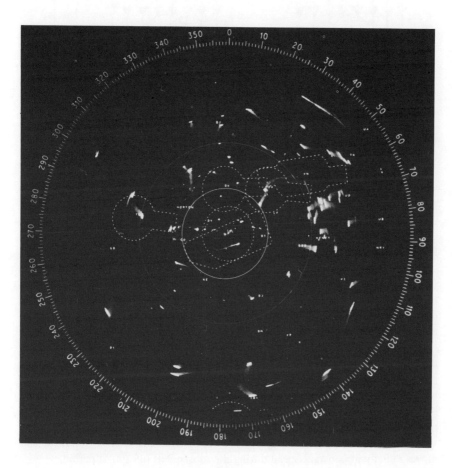

(b)

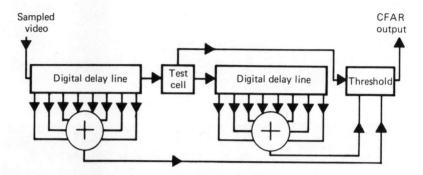

Figure 5.16   Cell-averaging CFAR

threshold less susceptible to individual target echoes entering the delay line. Conversely, fewer taps make the system more responsive to rapid changes in clutter level. If strong target echoes tend to spill over into cells adjacent to the test cell, the neighbouring cells can be omitted from the summation process. Averaging can also be based on azimuth as well as range cells.

5.7 Plot extraction and display

In this section we consider how radar information is prepared for presentation to the human operator. The main topics to be covered are analog and digital display techniques, the digital transmission of radar data over long distances, and modern developments in automatic detection and computer-driven displays.

We have already made brief comments about *plan-position-indicator (PPI)* displays in section 1.2.1. The PPI display format offers the great advantage of a natural, 'bird's-eye' view of the radar scene. In its conventional analog form it has a rotating trace which corresponds to the antenna beam azimuth. An incoming target echo causes bright-up of the trace at a radius proportional to slant range. The bright-up is achieved by intensity-modulation of a flying spot. The spot starts at the centre of the display each time the transmitter fires, and moves radially at constant speed during the subsequent interpulse period.

Figure 5.17 shows, in simplified form, a system for generating a rotating trace. An angle transducer (AT) is mounted on the shaft which connects the antenna to its motor–gearbox (M–G). The transducer output is proportional to antenna azimuth angle (θ). An associated *resolver* unit produces the signals $V\cos\theta$ and $V\sin\theta$ (V being a DC voltage), which are used as 'aiming voltages' for two sawtooth generators. The latter provide X and Y timebases for the PPI display. For example, the upper generator's output is a repetitive sawtooth waveform with a peak value $V\sin\theta$. The sawtooth begins when the transmitter fires, and resets at the end of the interpulse period. It is not difficult to see that

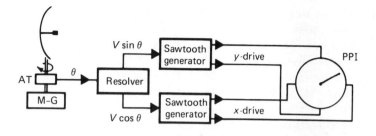

Figure 5.17 Generation of a rotating trace on a conventional PPI display

the varying amplitudes of the two sawtooth signals must produce a rotating trace on the screen.

In most modern systems the angle transducer is a digitally coded disc. A 12-bit (4096-step) code is quite common, giving an angular resolution better than $0.1°$.

Although the intensity-modulated PPI display is very widely used, various other display formats are encountered in radar. They include the *A-scope*, *B-scope* and *range-height indicator (RHI)*, defined as follows:

A-scope. An X–Y presentation in which the abscissa corresponds to range, and the ordinate corresponds to signal amplitude. Azimuth is not directly represented. An A-scope is similar to the usual type of oscilloscope display

B-scope. An intensity modulated X–Y presentation, with the abscissa corresponding to azimuth and the ordinate to range

RHI. An intensity-modulated X–Y presentation, with the abscissa corresponding to range and the ordinate to height

The conventional analog PPI display has an important drawback. The radar operator is required to work in semi-darkness. Basically this is because the display is driven directly by a radar video signal with a low refresh rate. Even when persistent screen phosphors with long afterglow are used, it is difficult to get enough energy into each 'blip' to make it reasonably bright. For example, a typical terminal area ATC radar may use a $1\ \mu s$ pulse length and achieve a nominal 15 hits-per-target at an antenna rotation speed of 15 rpm. Each blip must therefore be 'painted' on the screen in a total time of about $15\ \mu s$. But it must remain visible for at least 4 s. Such displays cannot cope with high ambient light conditions — for example, in airport control towers.

A special type of bright display known as the *direct-view storage tube (DVST)* was developed in the 1960s. Based on cathode-ray tube (CRT) technology, its principle was to use two separate electron beams. The so-called *flood beam* allowed a bright and persistent display of information carried by the *writing beam*. Although DVSTs found some acceptance with radar operators, they had poorer resolution than conventional CRT displays, with a tendency for gradual loss of picture quality in regions of fixed clutter.

Many modern bright displays employ the *scan-conversion* principle. The polar (range–azimuth) information of a conventional PPI is converted into the type of rectangular *raster* used by a TV monitor. The raster is stored digitally, and read out repetitively to produce a bright flicker-free display. Recent advances in raster-scan circuits for computer displays have made an important contribution in this area. Other advantages of raster-scan include the ability to combine signals from more than one radar on a single display, and the relative ease with which map data or alphanumerics can be inserted. Note, however, that radar applications normally demand high resolution. For example, the specification of one modern airport surface movement indicator (ASMI) radar includes a raster-scan TV display with more than 1000 horizontal lines (see figure 1.4).

We now turn to a rather different topic – the transmission of radar information from a radar site to a distant ATC or operations centre. As we have already noted, radar signals after detection and processing are essentially *video signals*. Typical bandwidths are between 250 KHz and 5 MHz. In this form they may be transmitted economically over modest distances by high-grade electrical cable or optical fibre. Alternatively, if the distance is great, or the terrain is difficult, they may be sent by microwave radio link. However, the modern approach is to extract the essential information at the radar site, and transmit it digitally via a standard telephone channel.

To understand how this is achieved, we need to consider the information content of a typical radar screen. Let us again take an ATC surveillance radar as an example. In normal air traffic conditions there might be 100 aircraft visible on a PPI display. Suppose that each aircraft position needs specifying in both range and azimuth (or in X and Y co-ordinates) with an accuracy of 1 part in 1024. Therefore we can describe each position using a 20-bit binary code (10 bits for range, 10 for azimuth), and transmit the complete radar picture using just $100 \times 20 = 2000$ bits. If we assume an antenna rotation speed of 15 rpm, the information needs updating every 4 s. The average bit rate is therefore 500 bits per second, or bauds.

We have, of course, taken a rather simplified view of the problem. Additional information may have to be sent – for example, weather data, antenna azimuth data and synchronising pulses. Extra control and parity bits will probably be needed. A secondary radar may be installed at the same site. Two-way communication may be required for remote monitoring and control. Nevertheless, our calculations show clearly that a standard telephone channel, working at a typical rate of 2400 bauds, should be quite capable of transmitting the radar information.

The reader may suspect a catch in the argument somewhere. How can a radar signal be passed along a telephone cable with a bandwidth of around 3 kHz? The answer is that all decisions about the presence or absence of targets, clutter and false alarms due to noise must now be made automatically at the radar site. The fine detail of a conventional analog display — detail which an experienced operator interprets with great skill — is removed. The radar operator becomes wholly dependent on automatic detection circuits and computers. As our discussion in the previous section makes clear, these must normally include some form of CFAR processing.

Automatic detection and coding of target data prior to digital transmission is widely referred to as *plot extraction*. In the early days of digital signal processing applied to radar, a separate plot extractor was often interfaced to existing equipment. Nowadays, however, plot extraction tends to be fully integrated into the signal processing chain. It may also be referred to as *data extraction*, or *target evaluation*. In the case of MTD systems, the term *hit processor* (see figure 5.14) is quite common.

The facilities provided by a plot extractor vary somewhat from one installation to another. The following stages are typical:

Grouping together all hits which appear to come from the same target
Eliminating false alarms due to noise, and unsatisfactory echoes deemed
    to come from clutter or angels
Estimating the range–azimuth (or X-Y) co-ordinates of all valid targets
Initiating, updating and terminating target tracks
Coding extracted data for digital transmission

Automatic prediction of future positions and velocities, and/or smoothing (filtering) of target tracks, may also be provided. Plot extraction is also highly relevant to secondary surveillance radar. We comment again on this in section 6.2.

The details of plot extraction algorithms are beyond the scope of this book. However, we should say a few words about the first of the above stages — namely, the grouping together of hits which appear to come from the same target. This relates to the basic statistical nature of radar echoes. Consider, for example, an ATC radar which operates with a nominal 15 hits-per-target ($n = 15$). Following digital processing (typically, MTI/MTD with CFAR), a hit is registered whenever the signal level in a resolution cell exceeds the cell threshold. It is now necessary to examine 15 adjacent azimuth cells, all at the same range, to establish whether the hit forms part of a composite target echo. Even a strong target will not, in general, produce a full sequence of 15 hits because of the random effects of noise, clutter and target cross-section fluctuations. Denoting a hit by a '1', and a miss by a '0', the typical sequences shown in table 5.3 might be obtained for strong, medium and weak targets.

Each sequence must be tested to decide whether or not it represents a wanted target. A common technique for doing this is to use a set of *sliding (or moving) windows*. At each range value, a window n azimuth cells wide is continuously swept through the surveillance area. The window contents are summed and examined once every interpulse period. Whenever the count reaches a pre-

Table 5.3

Target	Hits and misses														
Strong	0	1	0	1	1	1	1	1	1	1	1	1	1	0	1
Medium	0	0	1	0	1	0	1	1	1	0	1	1	0	0	1
Weak	0	0	0	1	0	0	1	1	1	1	0	0	0	0	0

↑

Beam centre

determined value (say 6 in the above example), a target 'leading edge' is declared. When the count falls again to below this value, a 'trailing edge' is declared. Target azimuth can be estimated from the leading and trailing edge positions. Needless to say, there are a number of possible variations on this basic theme.

In effect, a sliding window detector performs a *running integration* of target echoes in adjacent azimuth cells. Its action is analogous to the *video integration* performed by the screen phosphor of a conventional, analog PPI display. We have previously discussed this effect in section 2.2.2.

Schemes of this type are also known as *double-threshold detectors*. The *first threshold* corresponds to that of the individual resolution cell, which determines whether a hit has occurred. The *second threshold* is the value of sliding window count which must be reached for a valid target to be declared. If the second threshold is raised, it becomes less likely that a group of adjacent false alarms due to noise or clutter will be interpreted as a target. But the chance of missing genuine targets increases.

It is now time to consider the display of plot-extracted radar data. We confine ourselves here to some general remarks. Firstly, digital target plots arriving at an ATC or operations centre (often via telephone landline) may be stored, and read out at a relatively high refresh rate. This offers the great advantage of bright, flicker-free displays. The radar information does not strictly have to be presented in real-time. A delay of a second or two is normally acceptable, allowing substantial processing of incoming data before presentation. Special target and tracking symbols may be generated, and map or weather data displayed. Alphanumeric labels may be attached to targets, giving identification and height derived from secondary radar responses. The list of possible facilities on a modern synthetic display is almost endless.

The underlying technology is, of course, that of digital electronics and computers. It is quite usual to find one or more powerful minicomputers at the heart of a modern radar display system. Individual displays, linked via the central computers, often have considerable autonomy and incorporate their own computers or microprocessors. Readers who are interested in systems of this type will find further information in section 7.4.

There is considerable interest in the use of colour for radar displays. Recent technical advances in computer colour monitors, together with decreasing cost, make this a genuine option for many radar applications. For example, relatively inexpensive civil marine radars are available with daylight-viewing colour displays based on digital scan-conversion techniques. Such displays are attractive to the user and can aid feature recognition. However, their superiority over well-designed monochrome displays is the subject of continuing debate. It seems generally agreed that not more than four or five separate colours should be provided. These can be used in many ways — for example, to distinguish between wanted echoes, alphanumerics, weather and map data on a PPI display.

We end the section with some remarks about the radar operator. In a typical radar environment there are long periods of relative calm, punctuated by occasional crises demanding clear decisions. An earlier generation of radar operators, viewing poor-quality displays in darkened rooms, were particularly subject to the effects of stress and fatigue. Nevertheless, many of them became highly skilful

Figure 5.18   A modern synthetic PPI display. Note the hand-operated tracker ball used by the operator to control symbols on the display (photo courtesy of Plessey Radar Ltd)

at interpreting analog radar displays, and had a true 'feel' for the statistical nature of radar echoes. It is very difficult to replace such experience and mental flexibility by automatic detection and tracking circuits. On the other hand, the advent of bright, computer-driven displays has greatly reduced demands on the human decision-maker, and provided a more attractive working environment. There is perhaps one danger with the new technology. Radar operators may place too much faith in 'clean', clear, unambiguous radar displays, and fail to take account of the essentially statistical nature of radar detection.

6 Secondary Radar

6.1 The basis of secondary surveillance radar (SSR)

SSR, also known in the USA as *beacon radar*, has a number of similarities with primary surveillance radar. Both use a ground-based rotating antenna, with a pulse transmitter and receiver. Like its primary counterpart, the signals from an SSR normally undergo plot extraction at the radar site, followed by transmission by landline to an ATC or operations centre. Secondary radar antennas are often mounted above primary antennas (see figure 1.2), saving the cost of separate drive motors and gearboxes, and ensuring azimuthal alignment. Plot extraction may include correlation of primary and secondary radar signals, giving increased confidence about the position and identity of wanted targets.

Unlike primary radar, SSR is essentially a co-operative system. Its transmissions trigger coded responses from suitably equipped aircraft. (In the military radar environment such techniques are known as *interrogation, friend or foe (IFF)*.) SSR does not suffer from the clutter problems of primary radar. Furthermore, the peak power output requirements of the radar transmitter and aircraft *transponder* are quite modest, since the transmission of each is only 'one-way'.

In principle, an aircraft transponder can supply many different types of information to the air traffic controller. In practice, it is aircraft identification and height which are most commonly requested. They are displayed as alphanumerics on radar PPI displays, adjacent to the relevant aircraft tracks.

SSR therefore supplies valuable additional information, over and above the range and bearing data provided by a conventional primary radar. However, it has its disadvantages. It depends entirely on co-operation by friendly aircraft, which must be fitted with equipment conforming to internationally agreed standards. And although it avoids the clutter problems of primary radar, it is subject to various types of interference. We describe these, and recent technical developments designed to counter them, a little later.

The interrogation signal transmitted from an SSR to aircraft comprises three pulses, which modulate a 1030 MHz carrier. The pulses are illustrated in figure 6.1a. The information requested of aircraft depends on the spacing between pulses P_1 and P_3 (pulse P_2 is used for sidelobe suppression, as described below). If the P_1–P_3 spacing is 8 μs, the radar is asking for aircraft *identity*, and is said to be working in *mode A*. If the spacing is 21 μs, aircraft *height* is being requested, and the system is working in *mode C*. Various other modes are

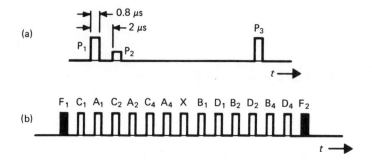

Figure 6.1 SSR interrogation and reply formats

available, but modes A and C are common to both civil and military aircraft and are the most widely used.

The form of reply transmitted by an aircraft transponder is shown in part (b) of the figure. All pulses are of 0.45 μs duration, with a 1 μs space between them, and modulate a 1090 MHz carrier. The *framing pulses*, F_1 and F_2, drawn solid in the figure, are always present. Data pulses A, B, C and D (three of each) provide a 12-bit binary code for reply information, and may or may not be present in a particular case. The central X pulse is not currently used. The 12-bit code provides up to 4096 separate aircraft identifications. Alternatively, it is used in mode C to report height in 100-feet increments, known as *flight levels*.

A conventional SSR puts out a continuous train of interrogations, at a rate comparable to the PRF of an en-route primary radar (say 400 per second). When, as is often the case, both aircraft identification and height are required, the interrogations may be *interlaced* (that is, alternated) between mode A and mode C.

As the antenna beam scans past a co-operating aircraft, a number of replies — say between 15 and 20 — are received. Like primary radar, target range may be accurately found from the time-delay between transmission and reception. Azimuth is more difficult, because the replies are spread over the horizontal beamwidth — typically 4 or 5 degrees — and individual replies may be missing or corrupted. The conventional approach has been to estimate azimuth using a *sliding window* technique, much as described in the previous section. Unfortunately, it does not yield very accurate results. The azimuth 'jitter' between successive scans is often around $0.15°$ rms. Air traffic controllers cannot be certain of an aircraft's location, or whether it is turning. As we shall see in the next section, azimuthal accuracy is greatly improved using the more up-to-date *monopulse* technique.

We have already mentioned that pulse P_2 shown in figure 6.1a is used for sidelobe suppression. SSR is much more susceptible than primary radar to the unwanted effects of antenna sidelobes. Relatively weak interrogating signals, reaching an aircraft via the sidelobes, may still be quite sufficient to trigger transponder replies at full strength. Azimuth measurement is further degraded, and the transponder is kept busy supplying unwanted replies. To prevent this happening, the SSR uses a directional main beam for transmitting pulses P_1 and P_3, and a very broad 'control' beam for transmitting P_2. We see this in figure 6.2. The gain of the control beam exceeds that of the main beam in all directions except that of the main lobe. A transponder receiving interrogations via a side-lobe detects a stronger P_2 pulse than a P_1 (or P_3) pulse. In the main beam itself, the reverse is true. Therefore, an amplitude discriminator in the transponder can determine whether the aircraft is in the main beam, and must reply; or in a side-lobe, where no reply is wanted.

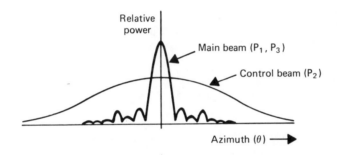

Figure 6.2 The use of main and control beams for sidelobe suppression

Now that we have described the coding of SSR transmissions and replies, we should consider the signal power levels involved. In chapter 2 we noted that very large pulse powers are required of long-range primary radar transmitters (assuming that pulse compression is not used). Furthermore, their receivers must be highly sensitive. We shall see that SSR is far less demanding, because the aircraft produces an active response.

We start by considering the path between SSR transmitter and aircraft, known as the *up-link*. If the transmitter power is P_t, then the power density reaching an aircraft at the centre of the main lobe is given by expression (2.2)

$$\frac{P_t G_1}{4\pi R^2} \tag{6.1}$$

where G_1 is the peak antenna gain and R is the range. Now suppose that the aircraft transponder has an antenna of effective area A_e and gain G_2. The received power is

$$P_r = \frac{P_t G_1 A_e}{4\pi R^2} \tag{6.2}$$

Equation (3.16) in chapter 3 expresses A_e in terms of antenna gain and the radar wavelength

$$A_e = \frac{G_2 \lambda^2}{4\pi} \tag{6.3}$$

Substitution gives

$$P_r = \frac{P_t G_1 G_2}{16\pi^2 R^2} \quad \text{or} \quad P_t = \frac{16\pi^2 R^2 P_r}{G_1 G_2 \lambda^2} \tag{6.4}$$

Let us now insert some typical values, to find the transmitter power required for interrogation at a range of 250 nm (460 km). We take $P_r = 10^{-10}$ W, which is broadly in line with the specification for transponder receivers issued by the International Civil Aviation Organisation (ICAO). The gain of a conventional SSR antenna is normally considerably less than that of a primary surveillance radar — say 23 dB (G_1 = 200). A transponder antenna is essentially omnidirectional in azimuth, being required to receive (and re-transmit) in any direction. In practice, G_2 is typically a few dB. However, if the aircraft banks the antenna may be partly shielded by wings or tail, and the effective gain is reduced. We therefore assume a value of 0 dB (G_2 = 1). Since the transmission frequency is 1030 MHz, λ = 29.1 cm. Hence

$$P_t = \frac{16\pi^2 \, (460 \times 10^3)^2 \times 10^{-10}}{200 \times 1 \times 0.291^2} \approx 200 \text{ W} \tag{6.5}$$

A realistic value would probably be nearer 500 W than 200 W, because our calculation has taken no account of such effects as transmission line losses or atmospheric attenuation. In any event, it is a great deal less than the megawatt pulse powers typical of long-range primary radar. 500 W peak pulse power corresponds to a few watts of mean power, so there are no significant transmitter cooling problems, and the electrical stresses are relatively low.

It is tempting to suggest that an SSR transmitter could work at much higher power, to allow interrogation of even more distant aircraft. However, we must remember that jet aircraft at normal cruising heights of around 30 000 feet (say 10 000 m) become invisible to a line-of-sight radar at ranges of about 250 nm, because of the Earth's curvature (see figure 3.5). It would in any case be undesirable to increase the range much above 200 nm, because aircraft transponders would spend a lot of their time replying to very distant SSRs which had little need of their information.

With these comments in mind, it is interesting to note that the ICAO specification for SSR transmissions includes a maximum permitted power level. Since the power density reaching an aircraft depends on the product of transmitted power output and antenna gain ($P_t G_1$), it is this product which is specified. It is called the *effective radiated power ERP* (strictly, P_t here denotes the actual power supplied to the antenna, taking transmission line losses into account). The maximum permitted ERP is 52.5 dBW (that is, dB with respect to 1 W). In our own example, we assumed an antenna gain of 23 dB, and calculated P_t = 200 W, or 23 dB above 1 W. Hence our estimated requirement for ERP was 46 dB. This puts the ICAO maximum figure into perspective.

The return transmission path from aircraft to SSR is called the *down-link*. Fortunately, the reciprocity property of antennas means that equations (6.1) to (6.4) again apply. Since the frequency shift between up-link and down-link is only about 6 per cent, the values of G_1 and G_2 are little changed. In other words, range performance in the two directions is essentially similar. The international specification for the down-link calls for transponder ERPs in the range 18.5–27 dBW, corresponding to peak power outputs between 70 W and 500 W for antennas with 0 dB gain.

We have not so far mentioned noise or signal statistics in our description of SSR. This contrasts with the space devoted to such matters in chapter 2. Although it would be wrong to imply that they are unimportant in SSR, they do not play the same dominant role as in primary radar. There are three main reasons. SSR does not suffer from clutter which is, of course, essentially statistical. Secondly, signal levels in SSR are designed to provide generous signal-to-noise ratios for detection of long-range aircraft. And finally, SSR is not affected by aircraft cross-section fluctuations. A transponder reply from a light aircraft is as strong as that from a large jet, and is little affected by changes in attitude.

The main problems of SSR are rather different. They fall into two principle categories: *co-channel interference*, and *multipath interference*. We now describe each of them in turn.

Co-channel interference arises from the use of up-link and down-link frequencies which are common to all SSRs and transponders. It may be divided into three types, known as *capture*, *fruit*, and *garble*.

Capture, or *lock-up*, occurs when an aircraft is already replying to one SSR station, and cannot therefore reply to another at the same time. Lost replies jeopardise the measurement of azimuth by the sliding window technique, and in severe cases cause unreliable data decoding.

Fruit ('false replies unsynchronised in time') denotes aircraft replies in response to interrogation by other SSR stations. Fruit appears as random, or quasi-random, interference. It, too, can affect azimuth measurement and the decoding of wanted replies.

Garble denotes the overlap of two (or more) separate aircraft replies. It occurs when aircraft are close together in range, and can cause considerable problems in detection and decoding.

Unfortunately all these types of interference are more frequent in busy air traffic conditions, which is just when reliable radar data is most needed. It is easy to appreciate that co-channel interference is made worse by the blanket nature of SSR transmissions. Each SSR interrogates all aircraft within range, and accepts all their replies. It also accepts replies triggered by other SSRs. Clearly, the situation would be improved by making the system more selective. We return to this important idea in section 6.3.

Multipath interference is caused by unwanted reflecting paths between SSR and aircraft. If the path difference between direct and reflected energy is small, the effect is to produce phase interference between the wanted and unwanted paths. An important example is the production of lobes and nulls in the vertical coverage pattern of an SSR, caused by ground-reflected energy. We have previously discussed this phenomenon in some detail for primary radar in section 3.2. As far as SSR is concerned, it tends to cause missed detections on aircraft flying at certain elevation angles.

If the path difference is long, two versions of the signal arrive at substantially different times. Interference between various pulses in the direct and reflected signals may cause a transponder to reply to sidelobe interrogation, or it may corrupt the reply code.

A somewhat different multipath effect is the generation of 'ghost' aircraft. If the main beam of an SSR points at a large reflecting object – for example, a hangar, parked aircraft or even a chain link fence – it may interrogate a target by reflection. A ghost aircraft then appears in the same direction as the reflecting object. Its azimuth may be quite different from that of the true target, leading to considerable confusion.

We see that conventional SSR suffers from a variety of complicated interference effects, which are hard to quantify. Such limitations were less important in the days when SSR was seen as an adjunct to primary radar, rather than as a navigational system in its own right. However, the attraction of SSR to ATC authorities has stimulated some major technical improvements in recent years. We describe these, and current developments, in the following sections.

6.2 Monopulse SSR

The *monopulse* technique overcomes many of the performance limitations of conventional SSR. The term monopulse implies that target azimuth can be accurately measured using a single transponder reply. Azimuth jitter is typically reduced to $0.03°$ rms or better – roughly a five-fold improvement over conventional systems. This is achieved with a special horizontal beam pattern giving much greater selectivity. In addition, the antenna used with a monopulse SSR has a relatively large vertical aperture. Vertical beamwidth is smaller, with a well-controlled lower cut-off. Problems caused by ground-reflected energy are therefore much reduced.

Another advantage of monopulse SSR is that it decodes individual transponder replies with a high degree of reliability. Unlike conventional SSR, it does not need many replies from an aircraft on each scan of the antenna. A lower PRF can therefore be used. There is a general reduction of SSR 'message traffic', and correspondingly less co-channel interference.

We start by describing the special horizontal beam pattern used by monopulse systems. The main and control beams of a conventional SSR were shown in figure 6.2, and the reader will recall that the control beam is used for sidelobe suppression. In monopulse SSR the main beam is generally known as the *sum (Σ) pattern*. A further beam, known as the *difference (Δ) pattern*, is also introduced. The reasons for these names will become clear a little later. The central portions of a typical sum and difference pattern are illustrated in figure 6.3a.

The detailed use of the patterns during transmission and reception is quite complicated, and varies somewhat between equipments. However, the basic principle is readily explained. Whereas the sum pattern has its peak along the antenna *boresight* ($\theta = 0°$ in the figure), the difference pattern goes through a deep null. It is therefore possible to estimate the *off-boresight azimuth (OBA)* of a transponder reply by comparing the amplitudes of signals obtained via the two patterns. Since the slope of the difference pattern to either side of the boresight is very steep, an accurate OBA measurement can be obtained from a single transponder reply. Furthermore, the difference pattern exhibits a 180° phase reversal at its null. So by comparing the phase of sum and difference signals, it is possible to establish whether the aircraft is to the left or right of centre. The OBA value is then added to the boresight azimuth obtained from an angle transducer on the antenna shaft.

Sum and difference patterns can both be obtained from the same antenna. The techniques involved are closely related to our discussion of phased arrays in section 4.2.2, and are summarised by figure 6.3b. At the top is shown a typical amplitude distribution (A) across the horizontal aperture, tapered to reduce sidelobe levels. The sum pattern is produced by arranging a uniform phase distribution across the aperture – labelled Σ in the lower part of the figure. A monopulse SSR antenna achieves this with a horizontal array of discrete elements, all fed in phase. The resulting far-field pattern is therefore essentially a simple

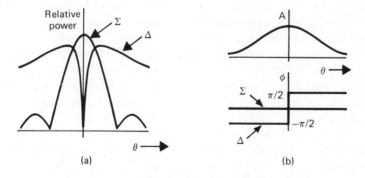

Figure 6.3 Antenna patterns for monopulse SSR

summation of amplitude contributions from the various elements. It displays the usual type of well-defined main lobe, centred on the antenna boresight.

The difference pattern is generated by introducing a phase reversal ($\pm \pi/2$) between horizontal elements to either side of the antenna centre-line (labelled Δ in the figure). In effect, the far-field pattern is now formed as the *difference* between amplitude contributions from the two sides of the antenna array. Fourier transformation of the aperture distribution, taking phase as well as amplitude into account, reveals that the far-field pattern now exhibits a deep null on the boresight.

Typical antennas of this type may have 30–40 horizontal elements, with apertures of 8 or 9 m. Referring back to table 4.1, we see that the 3 dB beamwidth of a tapered aperture distribution with uniform phase is around $70\lambda/a$ degrees (where a is the aperture). In current SSR systems, λ is close to 30 cm. These figures give a horizontal beamwidth for the sum pattern of about $2.5°$ – roughly half the value achieved with smaller, conventional SSR antennas. Of course, the effective *system* beamwidth is a lot less than $2.5°$, owing to the use of both sum and difference patterns.

We have previously noted that monopulse antennas also have a relatively large vertical aperture (LVA). A conventional ('hogtrough') SSR antenna might have a vertical aperture of 0.4 m, with a beamwidth of about $50°$. The vertical beam is ill-formed, with too much energy directed at high elevation angles and – more importantly – towards the ground. In LVA designs the aperture is more likely to be 1.5–2 m, producing a beamwidth of about $12°$. Typically, 10 or 12 rows of radiating elements are spaced across the vertical aperture. Careful amplitude and phase control gives a much more accurately shaped beam, with a sharp cut-off at low elevation angles. Multipath interference due to ground-reflected energy, which can have such an adverse effect on SSR performance, is much reduced.

Before leaving the subject of monopulse antennas, we might note that monopulse techniques are also very important in military applications. They are used in primary, as well as secondary, radar equipments. For example, phased arrays can produce monopulse vertical beam patterns, which greatly improve target resolution in elevation (and height). Although this large topic is beyond the scope of the present book, the reader should be aware of its significance.

Returning to our main theme, we must not give the impression that the success of civil monopulse SSR rests entirely on the design of a high-performance antenna. Some of the most striking recent developments are in signal processing and plot extraction. In view of our comments in chapter 5 on signal processing trends in primary radar, it is not surprising that computer techniques are nowadays very important in SSR.

The extraction of SSR plots involves a number of closely related stages. At some risk of oversimplification, we may summarise a typical processing sequence as follows.

Range, azimuth, and code recognition. The range and azimuth of individual transponder replies is estimated. Parallel processing hardware may be necessary

Figure 6.4   A large-vertical-aperture (LVA) antenna for secondary surveillance radar (photo courtesy of Cossor Electronics Ltd)

to cope with the high input data rates from sectors with dense air traffic. At this stage any garbled (overlapping) replies are separated, and valid codes are recognised.

Initial plot formation. Groups of replies from the same aircraft are associated by range and azimuth correlation, forming 'partial' plots or 'raw' target reports. Randomly occurring fruit is rejected, and gaps in reply sequences caused by capture or unsuccessful interrogation are bridged. This and subsequent processing stages do not strictly have to be carried out in real-time, although any computational delays should be less than (say) 1 second.

Final plot formation. The confidence associated with partial plots is increased by taking account of the track history of wanted targets, and of known reflecting objects likely to cause multipath errors. SSR plots may also be correlated with primary radar plots obtained at the same radar site. Final plots are typically reported in terms of range–azimuth (or X–Y) co-ordinates, reply code and interrogation mode.

There is clearly much scope for sophisticated computing algorithms. Well-designed algorithms make effective use of track histories, the high degree of correlation between successive replies from the same aircraft, and knowledge of local site conditions. As in primary radar (and especially MTD systems), signal processing is becoming increasingly adaptive. For example, unwanted 'ghost' replies caused by temporary reflecting objects – such as the tail fins of parked aircraft – can be recognised and rejected.

Extracted SSR plots are finally transmitted in serial digital form to an ATC or operations centre – often by telephone landline. The underlying principles have already been discussed for primary radar in section 5.7. In most modern systems the SSR information updates a computer store, and is subsequently read out at a relatively high refresh rate to bright radar displays.

The improved operational performance of monopulse SSR is leading many ATC authorities to re-examine the need for long-range primary radar. The reader will find additional comments on independent monopulse systems for en-route surveillance of civil air traffic in sections 7.3 and 7.5.

6.3 Mode-S operation

We have seen how conventional and monopulse SSR provide elementary two-way communication between an aircraft and the ground. In the Mode-S system, planned for the 1990s and beyond, the concept of a digital data-link is greatly extended. This is achieved by specifying much longer pulse trains for both up-link and down-link transmissions.

The second major feature of Mode-S is that it minimises broadcast interrogations. The 'S' in Mode-S signifies *selective*. Once an aircraft has been located, or *acquired*, the Mode-S system interrogates it only when the antenna is pointing more or less in the right direction. Furthermore, the ground station can choose which aircraft it wishes to interrogate. The *selectable-address* capability of Mode-S offers a major overall reduction of SSR message traffic, and hence in errors caused by capture, fruit and garble.

Nevertheless, in the lengthy changeover period between conventional/monopulse SSR and Mode-S, ground stations must be able to interrogate both types of transponder. The Mode-S system therefore employs the same up-link and down-link frequencies and uses monopulse antenna patterns. Current monopulse SSR equipments are generally designed for Mode-S compatability.

The message formats for Mode-S are shown in figure 6.5. Up-link transmissions are of two types. Part (a) of the figure shows the *all-call* format. It is used to acquire aircraft for Mode-S, and also to broadcast interrogations for standard SSR transponders. There is an extra pulse P_4 in the train, compared with figure 6.1. If P_4 is 1.6 μs long, Mode-S transponders reply with their own special code. But if P_4 is 0.8 μs long, they remain silent, and only conventional transponders reply.

Having acquired an aircraft, the Mode-S ground station knows its identity and approximate location. On subsequent antenna rotations it interrogates selectively with the pulse pattern shown in figure 6.5b. Pulses P_1 and P_2 have the same amplitude and are transmitted by the main beam. Since P_1 is not greater than

P_2, standard SSR transponders do not reply. An additional pulse, P_5 (not shown in the figure), is used for sidelobe suppression in Mode-S, and is transmitted by the control beam.

The main information content of a Mode-S interrogation is contained in pulse P_6, which comprises either a 56-bit or a 112-bit word. Individual bits are 0.25 μs long and are coded using *differential phase-shift keying (DPSK)*. Both '1's and '0's in the pattern are represented by the presence of a signal, with a 180° phase reversal whenever a '1' occurs. This type of modulation possesses high noise immunity. The first 1.25 μs of P_6 does not carry information bits, but is used to provide the phase reference needed by the transponder for phase-sensitive detection.

The Mode-S transponder reply format is shown in part (c) of the figure. Following a four-pulse control sequence, *or preamble*, the main data block again contains either 56 or 112 bits. In this case a form of pulse-position coding is used. Within each 1 μs time slot there is a 0.5 μs pulse. If the pulse occupies the first half of the slot, it denotes a '1'; if it occupies the second half, it denotes a '0'.

The information capacity of Mode-S interrogations and replies is clearly much greater than that of conventional or monopulse SSR. Indeed, the international specification allows it to be extended even further by stringing together a number of transmissions to form *extended-length messages*. In principle, Mode-S SSR can be used to carry such information as:

ATC flight clearance data
Airborne collision avoidance and minimum safe altitude warnings
Automatic reporting of aircraft flight parameters
Weather reports.

Much of this data has conventionally been exchanged between pilots and air traffic controllers using voice communication.

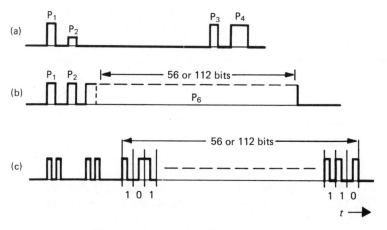

Figure 6.5 Mode-S message formats

7 Modern Surveillance Radar for Civil Air Traffic Control

7.1 Introduction

This chapter gives a brief account of civil ATC radar in Britain, France, Germany and the USA. All four countries have high-density air traffic, and their ATC authorities operate comprehensive primary and secondary radar facilities. Radar is continuously evolving, and the systems described here are no exception. Nevertheless, our account should give the reader an overall view of current equipment and trends. To save space and avoid repetition, we focus on somewhat different aspects of ATC radar in each country.

Before discussing details, it is helpful to make a few general points:

Primary and secondary radar. SSR is now firmly established as an important element in international ATC. The capital cost to aviation authorities is considerably less than for long-range primary radar. In general, there is a continuing requirement for both primary and secondary radar coverage of civilian airspace.

Signal processing. There is a clear trend towards MTD-style processing and digital plot extraction in primary ATC radar, together with pulse compression for long-range systems. Substantial improvements in SSR performance are offered by monopulse and Mode-S equipments.

Remote/unattended operation. Ground-based radars, especially for en-route surveillance, are often sited well away from airports and ATC centres. Operating costs are reduced if equipment can be left unattended. This trend is supported by two main technical developments. The first is improved reliability, as a result largely of digital techniques, and the use of pulse compression in primary radar. Secondly, the increasing use of built-in test equipment (BITE) allows any faults to be reported back automatically to the appropriate maintenance centre.

7.2 Britain

Our discussion of ATC radar in Britain concentrates on a series of surveillance radars installed by the British Civil Aviation Authority during the 1980s. The primary radar equipment (Hollandse Signaalapparaten) incorporates many advanced features such as pulse compression and full MTD-style signal processing. Digital radar signals are integrated with those obtained from co-located SSRs (Cossor) before transmission to the London and Scottish ATC Centres. The primary antenna (AEG) and associated SSR antenna (Marconi) of one such radar have already been illustrated in figure 1.2.

The locations of the radars are shown in figure 7.1a. (There are, in addition, some 15 surveillance radars of various types installed at or near regional airports, providing local terminal-area/approach-control services.) The system at London Heathrow (LH) is designed for approach control, plus terminal-area surveillance for the London ATC Centre, and has a nominal range of 80 nm. The radars near Gatwick (G) and Debden (D) provide approach-control services for London Gatwick and Stansted airports respectively, together with long-range surveillance and supplementary terminal-area surveillance for the London ATC Centre. Their nominal range is 160 nm. The equipments at Claxby (C), Great Dun Fell (GDF), and the Isle of Tiree (T) are for long-range surveillance, with a displayed range of 250 nm. Finally, there is an additional, transportable radar at the Technical Support Facility near Gatwick. It is used for testing and evaluation, and acts as a standby system.

The equipment configuration at each radar station is illustrated in part (b) of the figure. It comprises fully duplicated primary radars (PR) and monopulse secondary radars (SSR). The only major common item is the antenna assembly, with its rotating joints (RJ), primary radar antenna (PA) and LVA secondary antenna (SA). Both signal processors (SP) integrate the returns from the two PR

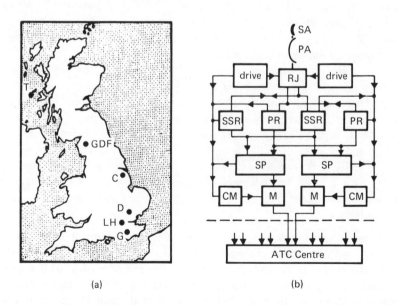

(a)   (b)

Figure 7.1   Major ATC surveillance radars in Britain: (a) locations and (b) equipment configuration

and SSR equipments, forming digital plots which are transmitted to ATC centres via modems (M) and landlines. (Analog radar signals are also available for use by local approach control services.) Extensive remote control and monitoring of equipment status is provided by dual control/monitoring (CM) systems (Marconi). In spite of the rather different operational roles of the various radars, all have the same equipment configuration. However, antenna rotation speeds vary from 7.5 rpm for long-range surveillance to 15 rpm for approach control (London Heathrow).

We now focus on the main technical features of the primary radar equipments. The parameter values quoted are typical, rather than definitive. Leaving for the moment the question of signal processing, the features may be summarised as follows:

Antenna. The L-band modified-parabolic antenna measures 14.5 m × 9 m and produces a horizontal beamwidth of just over 1°. Dual horn feeds give a low pencil beam for long-range coverage and a higher fan beam for closer targets. Both horns are 'active', the transmitter power being fed to pencil and fan beams in the ratio 1:3. On receive, beam selection is controlled in range and azimuth according to the required vertical coverage and local clutter conditions.

Transmitter. Transmission is at about 1300 MHz in frequency diversity, with a typical separation of 60 MHz between the two channels. The power output stages are TWTs with a nominal peak power of 150 kW. FM pulse compression is used. The main transmitter pulse, known as the *long-pulse*, is of 66 μs duration, and is compressed to 0.6 μs in the receiver. Additional *short-pulses* of 3 μs duration are used to detect targets at close range (< 6 nm), and are transmitted between successive long pulses. They, too, are compressed to 0.6 μs, giving a system range resolution of about 90 m.

Receiver. Each duplicated receiver has three channels: one for long-pulse echoes received from the pencil beam; one for long-pulse echoes received from the fan beam; and the third for short-pulse echoes received from either beam. Beam selection is achieved by switching between the channels. A sophisticated form of STC is provided for each channel, adjustable in 128 independent range intervals and 32 azimuth sectors according to local clutter conditions. Low-noise preamplifiers give typical noise figures of about 3 dB. Following pulse compression by SAW devices, and coherent phase-sensitive detection, all three receiver channels produce I and Q signals for subsequent digital signal processing. 12-bit coding is used.

The overall signal processing scheme is illustrated by figure 7.2. Outputs from the two long-pulse receiver channels are passed to MTD systems similar to that already shown in figure 5.14. A single canceller precedes each MTD doppler filter bank, giving an extra 20 dB discrimination against fixed clutter. Doppler filtering is based on 16-point FFT hardware, with each interpulse interval divided into range increments of 1/32 nm. The zero-velocity filter is a 16-tap

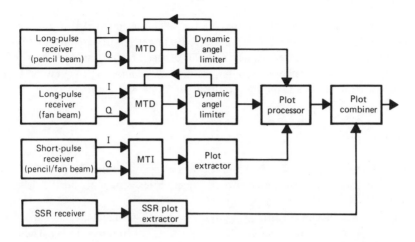

Figure 7.2 Signal processing scheme of the radars shown in figure 7.1

transversal filter, and the associated clutter map has clutter cells of size ½ nm ×
1.4°.

The output from each MTD hit processor is passed to an additional processing
stage known as a *dynamic angel limiter (DAL)*. This restricts the total number of
targets which can be reported, taking into account recent traffic density in a
given sector and the overall plot energy. If the number of targets rises substantially above the expected value — owing, for example, to a flock of birds —
the DAL adjusts the relevant MTD thresholds to prevent system overload.

Two further features of the MTD scheme deserve mention. The ambiguities
in target radial velocity inherent in MTD processing are resolved in this system by
operating the dual transmitter/receivers in frequency diversity (rather than by
altering the PRF, as mentioned in section 5.5.1). The radial velocity scales are
therefore different in the two receivers, and moving targets appear at different
filter outputs. The second feature is the provision of weather data. The outputs
of FFT filters 2–14 are passed to a weather detector, which uses two thresholds
to distinguish 'moderate' and 'severe' weather. The resulting weather map indicates the range limits of any such weather in each 2.8° azimuth sector, and can
be displayed to air traffic controllers.

Figure 7.2 shows that signals from the short-pulse receiver channel are processed by an MTI, rather than an MTD, system. This is because several short
pulses are transmitted for every long pulse, giving a higher PRF and more hits-
per-target in the short-pulse channel. Therefore, the rather simpler MTI approach
to clutter cancellation proves quite adequate. A four-pulse canceller is used, and
the same piece of hardware is used to provide a zero-velocity filter. In other
respects the processing is very similar to that of the MTDs.

The plot extractor at the output of the MTI system works on the sliding-window principle described in section 5.7. Its target reports are integrated with those from the long-pulse channels in a *plot processor*. The final task is to corre-late primary and secondary radar targets in a *plot combiner*, for onward trans-mission via modems and landlines to the appropriate ATC Centre.

7.3 France

In this section we describe the main features of primary surveillance radars used by the French ATC Authority. We also make brief comments on the develop-ment of SSR coverage in France.

The primary radars fall into three groups: en-route, terminal area and approach control (Thomson/CSF types LP 23, TR 23 and TA 10 respectively). There are some 14 radars in all, at the locations shown in figure 7.3. An additional en-route radar, close to the Swiss border at La Dole, is operated by the Swiss Aviation Authority. The two main Paris airports (Charles de Gaulle and Orly) each have a terminal area radar and an approach control radar.

All these radars employ modified parabolic antennas and (with one exception) conventional magnetron transmitters. It is instructive to compare the three types

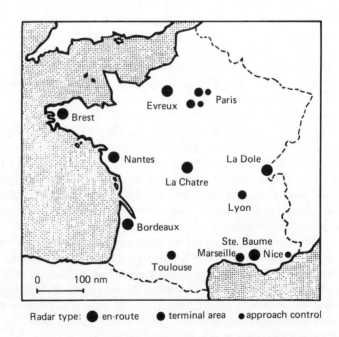

Figure 7.3 ATC surveillance radar sites in France

of system, to illustrate design choices associated with their different operational roles. Table 7.1 gives nominal values for the major technical parameters.

Table 7.1

	En-route	*Terminal area*	*Approach control*
Radar band	*L* (23 cm)	*L* (23 cm)	*S* (10 cm)
Displayed range	180 nm	120 nm	60 nm
Peak power	2.2 MW	2.2 MW	0.75 MW
Pulse width	3 μs	1.5 μs	1 μs
PRF (staggered)	360 Hz	700 Hz	1000 Hz
Antenna size	13 m × 7 m	9 m × 5 m	4.7 m × 2.3 m
Antenna gain	35 dB	33 dB	34 dB
Rotation rate	6 rpm	7.5 or 15 rpm	15 rpm
Horizontal beamwidth	1.1°	1.7°	1.5°

Information courtesy of Thomson/CSF.

The en-route systems include digital signal processing and plot extraction at the radar site, followed by data transmission over landlines to the appropriate ATC centres. At Bordeaux, Brest and Ste. Baume the radar is relatively close to the ATC Centre, allowing analog radar signals to be displayed as well as synthetic tracks.

The radars use dual transmitter/receivers working in frequency diversity, dual-channel (*I* and *Q*) digital MTI and PRF stagger. Stacked beams provide high and low vertical cover, switched according to the local clutter environment. An advanced form of STC is used. This adapts itself to fixed clutter with the aid of a continuously updated clutter map. Although the precise technical facilities of the radars vary with site and date of installation, recent models incorporate such features as built-in test equipment (BITE), remote monitoring and automatic system reconfiguration. MTD-style processing is also available.

The L-band primary radars described above have co-mounted SSR antennas. However, the French ATC authorities are replacing their conventional SSRs by about 20 monopulse equipments in the period to 1992. Designed for unattended operation, they include remote monitoring and control. Assuming their performance meets expectations, the en-route primary surveillance radars are likely to be phased out. In the longer term, the SSR network will be upgraded to Mode-S operation.

7.4 Germany

Our description of the German radar scene concentrates on long-range primary radar, and on the advanced computer-controlled radar display system used in a number of regional ATC centres.

Primary radar provision in the German Federal Republic (West Germany) includes 3 long-range systems sited near Frankfurt, Hanover and Munich, plus some 15 shorter-range radars based at or near major airports. Most have co-mounted SSR antennas. The long-range primary radars (AEG) are coherent-MTI systems with typical main parameters as follows:

antenna
modified parabolic, 14.5 m × 9 m
horizontal beamwidth: 1.05°
gain: 36 dB
sidelobe level: −24 dB
rotation rate: 5 rpm
polarisation: vertical or circular

transmitter
frequency: 1250–1350 MHz
peak power (klystron): 2.5 MW
pulse length: 2 μs
PRF in range: 310–480 Hz

receiver
noise figure: < 2.5 dB
IF: 30 MHz
IF bandwidth: 780 KHz
normal radar video: logarithmic

MTI
digital, dual-channel (I and Q)
ADC: 10-bit resolution
double cancellation with PRF stagger

The quoted maximum range is about 200 nm on a 5 m^2 target, with a detection probability P_d = 0.8 and a false-alarm probability P_{fa} = 10^{-6}. Dual transmitter-receivers are normally worked in frequency diversity, with a transmitter frequency separation of at least 40 MHz, and a timing separation of 4 μs between pulses on the two channels.

The antenna has two beams: a high beam giving cosecant-squared cover and a near-horizontal pencil beam for enhanced detection of long-range aircraft. On reception, the relative contributions from the two beams may be digitally controlled in four steps. In this way the overall vertical coverage pattern can be adjusted to take account of the severity of ground clutter, as a function of azimuth.

Clutter echoes are further reduced by an advanced form of STC, pre-programmable according to range and azimuth sector, and continuously adapting itself to take account of strong local clutter echoes. After MTI processing the primary radar signals are correlated with SSR returns in plot-extraction equipment, and the data sent via modems and landlines to ATC centres.

In the remainder of this section, we focus on the display of plot-extracted primary and secondary radar data in ATC centres. The Federal Air-Safety Authority (Bundesanstalt für Flugsicherung) operates 4 Regional Air Navigation Service Units (RANSUs) located at Bremen, Dusseldorf, Frankfurt and Munich. They are responsible for both en-route and terminal-area radar control. In addition 6 Local Air Navigation Service Units (LANSUs) are responsible for terminal-area control only (Cologne–Bonn, Frankfurt, Hamburg, Hanover, Nuremberg and Stuttgart). At all these units the radar and other data is presented to air traffic controllers on advanced computer-controlled synthetic displays.

The basic architecture of the so-called DERD–MC system (Display of Extracted Radar Data–Minicomputer) involves redundant minicomputers for radar pre-processing and system control. These main computers communicate with semi-autonomous radar displays. The equipment in all the ATC service units is broadly similar, although the RANSUs have somewhat more powerful minicomputers plus an additional back-up channel, and can accommodate a larger number of radar displays. The overall scheme in a typical RANSU is illustrated in very simplified form by figure 7.4.

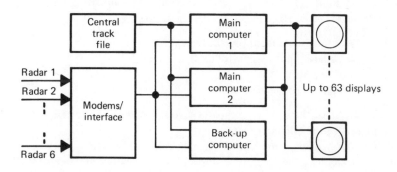

Figure 7.4   The DERD–MC radar display system used by the German ATC service

The system accepts plot-extracted data from up to 6 radars. It can drive up to 63 displays. The main minicomputers (Siemens) have the following major tasks:

Location of incoming radar data in a stereographically corrected master co-ordinate plane

Maintenance of a central track file
Processing of weather radar data
Storage of master maps (showing airways, control zones, etc.)
Overall system control and monitoring
Data recording
SSR code assignment

Most of these functions are also performed by the back-up computer.

Two types of radar display are provided (Raytheon, USA). One is for controlled lighting environments, and operates either in full-synthetic mode, or in mixed synthetic/analog mode. The latter is suitable for approach control, where it is often desirable to show 'raw' radar echoes together with digital target labels. The second display type is for control tower environments. It has an additional raster scan mode for coping with high ambient light levels.

As well as being in two-way communication with the main computers, each display unit incorporates a powerful minicomputer (Raytheon). This allows it to control and refresh the screen, and to select and process a wide variety of display formats – for example, various display scales and origins, maps, weather data, direction-finder (DF) inputs, target symbols and 'tails'. Control is by typewriter-style keyboard and tracker ball.

The main computers run identical software. The changeover strategy in case of failure is based on monitoring by all the radar displays. If a pre-determined proportion of displays consider that the main channel computer presently in use is faulty, the standby channel takes over.

It is clearly important that computer processing does not produce unacceptable time-delays. The DERD–MC system has the following typical response times:

Display of radar data from time of receipt: 0.7 s
Local keyboard input by controller: 0.1–0.3 s
Display of new map: 4 s
Main computer changeover: 3 s

Even the longest of these times compares favourably with the refresh rate of a typical long-range surveillance radar.

Another key operational factor is reliability. The redundancy of the DERD–MC system was designed to ensure a mean-time-between-failures (MTBF) of at least 10 000 hours for the dual-channel minicomputer system, or 50 000 hours including the back-up computer. In the first 2 years of operation ($>$ 17 000 hours), no such failures were reported at any of the ATC centres. Individual radar displays, designed to achieve an MTBF greater than 600 hours, in fact gave a figure approaching 5000 hours.

7.5 United States of America

Our main aim in this section is to describe the latest generation of terminal area/approach control radars designed for use at civil airports in the USA. We also make some general comments on the planned development of ATC radar services, including SSR, up to the year 2000.

The Federal Aviation Administration (FAA) is installing about 100 new airport surveillance radars (ASRs) in the period to 1995. Known as the ASR-9, the system adopted (Westinghouse Electric) includes the following features:

> Dual S-band transmitter/receivers with klystron power amplifiers and coherent signal processing
>
> Upper and lower antenna beams with linear/circular polarisation and independent STC
>
> Full digital MTD
>
> Separate weather radar channel
>
> Co-mounted SSR, Mode-S compatible, with integrated processing of primary and secondary radar targets
>
> Unattended operation, remote monitoring/control, automatic testing and fault indication

The vertical coverage given by the lower and upper antenna beams 'in-the-clear' is shown in figure 7.5. The patterns are relevant to a small aircraft target ($\sigma = 1\ \text{m}^2$) with Swerling case 1 characteristics, detection probability $P_d = 0.8$, and false-alarm probability $P_{fa} = 10^{-6}$. They incorporate the effects of typical STC characteristics. An enhanced version of the radar, the ASR-9I, offers an extended range up to 100 nm.

The lower beam is used for transmission and reception, the upper beam for reception only. Although transmission is restricted to the lower beam, sufficient power is radiated at all elevation angles of interest. The upper beam is much less subject to ground clutter and is used for detecting short-range aircraft.

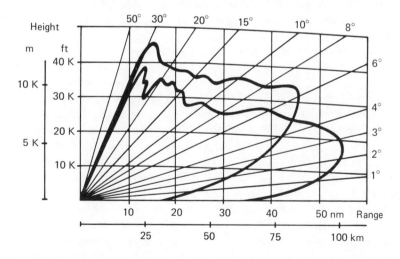

Figure 7.5 Vertical coverage patterns of the ASR-9 airport surveillance radar

The radar's range requirements can be met using conventional short pulses of moderate peak power. This simplifies receiver design, and avoids the problems of short-range detection posed by pulse compression techniques.

The signal processing system is designed to cope with small aircraft which may be flying tangentially to the radar in the presence of strong ground and weather clutter. Processing is based on I and Q signals sampled at 1.3 MHz with 12-bit resolution. The MTD scheme is broadly as shown in figure 5.14. It includes a doppler filter bank with programmable filter coefficients, zero velocity filter, and clutter map with 500 000 cells. Adaptive thresholding is achieved with a cell-averaging CFAR algorithm which uses a 'greater of' criterion. The algorithm is designed to ignore neighbouring targets in the cell-averaging process, which otherwise tend to raise the threshold and reduce detection probability.

Another interesting feature of the ASR-9 is the ability to desensitise, or 'sensor', particular areas in which unwanted moving targets are very likely to appear. Typically, these are stretches of highway visible from the radar site.

The MTD system divides the surveillance area into 256 azimuth sectors of about 1.4°, and processes them in turn. To prevent the loss of moving targets which appear at the output of doppler filter 0, each 1.4° sector is further divided into two equal *coherent processing intervals (CPIs)*. The PRF is fixed within any one CPI, but changes from one CPI to the next (in the ratio 7:9). Each CPI spans either 8 or 10 transmitter pulses.

The CFAR algorithm mentioned above allows good range resolution of adjacent targets with the same azimuth and doppler shift. A further signal processing algorithm improves the azimuth resolution of targets with the same range and doppler. If a train of echoes occupies more than two azimuth beamwidths, their amplitudes are compared with those expected from a single large target on the basis of the scanning modulation effect (see section 5.4.5). An amplitude sequence which deviates significantly from the expected one is interpreted as two separate targets.

The above stages of MTD processing may result in over thirty thousand threshold crossings on each revolution of the antenna. The actual number depends on air traffic density, clutter conditions and the presence of birds or angels. An individual aircraft may produce up to 50 crossings in different range, azimuth and doppler cells. Further processing stages are clearly needed to correlate all such 'primitive reports', eliminating unwanted echoes and estimating the positions and velocities of wanted targets. We have previously outlined such processing functions in sections 5.5.2 and 5.7, using the term *hit processor* to represent them in figure 5.14. In section 7.2 the terms *dynamic angel limiter* and *plot processor* were introduced to describe closely related operations.

In the case of the ASR-9, the additional processing is referred to as *correlation and interpolation (C & I)*, followed by *scan-to-scan correlation/tracking (SSCT)*. The first of these stages groups together primitive reports which appear to come from the same target, and resolves close-spaced targets. Range, azimuth, radial velocity and amplitude are extracted. In addition, each target is associated with a quality factor indicating azimuthal reliability. This stage of processing reduces false alarms to fewer than 60 per scan. The SSCT stage makes use of known scan-to-scan history to track moving aircraft, eliminating dubious or uninteresting targets. The false alarm rate is typically reduced to 1 per scan. Final target reports are associated with SSR replies before digital transmission by landline to the appropriate ATC centre.

The ASR-9 also features weather detection. It is used to indicate potentially dangerous weather to air traffic controllers. Two-level or six-level data can be supplied. The former is derived from the normal MTD system, by smoothing and superimposing output power levels obtained from the various doppler filters. Six-level data, which conforms to standards laid down by the American National Weather Service, is provided by a special weather channel. The channel allows for the selection of either linear or circular polarisation, and distinguishes between weather and ground clutter with the aid of a clutter map and special filters. The extracted weather data is transmitted to the ATC centre, where each controller can choose two weather contour levels for display.

Figure 7.6   ASR-9 antenna, with co-mounted SSR (photo courtesy of
Westinghouse Electric Corporation)

We end this section with some comments on the development of ATC radar services in mainland USA up to the year 2000. One of the most striking aspects is the scale of radar provision. In part this is due to the scale of the country itself. But it also reflects the growing importance of air travel, both public and private.

Figure 7.7 shows the locations of over 300 surveillance radars intended for the year 2000, as specified in a recent edition of the FAA's *National Airspace System Plan*. The total includes about 100 long-range Air Route Surveillance Radars (ARSRs) and about 200 medium-range Airport Surveillance Radars (including the ASR-9s), the great majority with co-mounted SSRs. There are also a number of SSR-only sites. About 200 SSRs are scheduled for conversion to Mode-S operation by the end of the century. Some of the present long-range radars (type ARSR-3) are likely to be decommissioned as Mode-S coverage improves. Others will be replaced by a new system, the ARSR-4.

The plan also increases the number of Airport Surface Detection Equipment (ASDE) radars to 30 (such radars are generally known in Britain as Airport Surface Movement Indicators, or ASMIs). A new generation of weather radars, together with the weather detection facilities included in the surveillance radars, will provide a total of about 130 weather radar locations.

Apart from the radars themselves, there are major developments in associated computer systems. The aim is to provide an integrated and highly automated national ATC service, based on about 20 Area Control Facilities. The traditional distinction between en-route and terminal area control will largely disappear. New software will aid prediction of traffic congestion and delays, leading to better overall safety and traffic flow management.

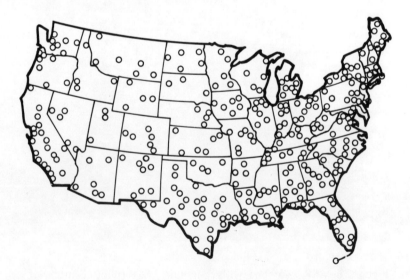

Figure 7.7 Planned ATC surveillance radar sites in the USA in the year 2000

Bibliography

In this short list of references for further reading, entries 1 to 5 are books, the rest are technical papers. Skolnik's comprehensive work of nearly 600 pages (entry 1) has an extensive list of references; however, it does not deal with secondary radar. Hovanessian's book of some 400 pages (entry 2) includes questions and worked examples. The book by Cole (entry 3), although written primarily for the radar user, gives an interesting practical account of secondary radar. Entries 4 and 5 are for the more specialist reader.

Among the technical papers, entries 6, 7, and 8 are important historically, and their authors' names have been mentioned in the main text. Entries 9 and 10 describe the new generation of civil ATC radars in the UK and USA respectively. Entry 11 covers technical and operational aspects of secondary surveillance radar.

1. Skolnik, M. I. (1980). *Introduction to Radar Systems*, 2nd edn, McGraw-Hill Kogakusha, Tokyo.
2. Hovanessian, S. A. (1984). *Radar System Design and Analysis*, Artech House, New York.
3. Cole, H. W. (1985). *Understanding Radar*, Collins, London.
4. Clarke, J. (Ed.) (1985). *Advances in Radar Techniques*, Peter Peregrinus, London.
5. Farina, A. and Studer, F. A. (1985). *Radar Data Processing*, Wiley, London.
6. Rice, S. O. (1945). 'Mathematical analysis of random noise', *Bell System Tech. J.*, **24**, 46–156.
7. Barlow, E. J. (1949). 'Doppler Radar', *Proc. IRE*, **37**, 340–355.
8. Swerling, P. (1960). 'Probability of detection for fluctuating targets', *IRE Trans.*, **IT-6**, 269–308.
9. Stokhof, W. (1981). 'Radars for an ATC system in the United Kingdom', *Philips Telecomm. Review*, **39**, 157–177.
10. Taylor, J. W. and Brunins, G. (1985). 'Design of a new Airport Surveillance Radar', *Proc. IEE*, **Special Issue on Radar**, 284–289.
11. Lewsey, R. (1986). 'Computer techniques for air traffic control radar', *Electronics and Power*, **32**, 309–311.

Problems

Section 2.1

Q1. A long-range surveillance radar is designed to detect a large aircraft ($\sigma = 10$ m^2) at 200 nm range. If sensitivity time control (STC) is not used, at what range in km would you expect it to detect a bird of effective area 5 cm^2?

Q2. A terminal-area radar delivers a peak pulse power of 500 kW. A satisfactory detection performance is obtained on a 1 m^2 target at 50 nm range if the received pulse power exceeds 10^{-13} W. What product of antenna gain (G) and effective receiving area (A_e) does the simple form of the Radar Equation suggest is needed?

Section 2.2.1

Q3. The average cross-section of a radar target is 1 m^2. Assuming it displays Swerling case 1 characteristics, find the probability that its cross-section is (a) less than 0.3 m^2, (b) greater than 3 m^2, on any one scan of the antenna.

Q4. Repeat problem Q3 for a Swerling case 3 target.

Section 2.2.2

Q5. The noise at the input of a radar IF amplifier is gaussian, with zero mean and a standard deviation of 100 mV.

What is the probability of finding the noise voltage within a 1 mV slot centred on (a) 100 mV, and (b) 500 mV?

Following envelope detection, the upper envelope has a Rayleigh pdf with $\psi_0^{\frac{1}{2}} = 100$ mV. What is the probability of finding this envelope within a 1 mV slot centred on (c) 100 mV, and (d) 500 mV?

If a detection threshold is set at 500 mV, find the false-alarm probability P_{fa}. Also estimate the average time between false alarms if the receiver bandwidth is 0.75 MHz.

Section 2.2.4

Q6. A long-range surveillance radar has a peak transmitter power of 2 MW, with a pulse length of 2.5 μs and a PRF of 400 pps. The antenna rotates at

5 rpm, has a horizontal beamwidth of 1.2°, and a gain of 34 dB. Its physical aperture is 13 m × 7 m. Additional parameters are as follows:

antenna efficiency ρ_A = 0.8
integration efficiency $E_i(n)$ = 0.65
noise figure F_n = 2.5 dB
IF bandwidth B = 400 kHz
system losses L_s = 5 dB

Estimate the maximum range in-the-clear on a 1 m² target with Swerling case 1 characteristics, for detection probability P_d = 0.8, and false-alarm probability P_{fa} = 10^{-6}.

Q7. An alternative version of the radar system in problem **Q6** employs pulse compression. 60 μs pulses are transmitted, followed by compression to 1 μs in the receiver. The IF bandwidth is increased to 1 MHz, and the noise figure is improved to 2 dB. The average transmitter power is unchanged. Estimate the change in range performance if all other parameters remain as before.

Q8. A range calculation for a small marine radar was given at the end of section 2.2.4 in the main text. By how many dB would its receiver noise figure have to be improved to give the same range performance, but on a Swerling case 1 target with P_d = 0.8 and P_{fa} = 10^{-6}?

Section 3.1

Q9. An X-band radar uses horizontal polarisation, with a horizontal beamwidth of 1°, and a pulse length of 0.5 μs. Echoes from a 1 m² target compete with heavy land clutter in the same radar resolution cell. The clutter comes from cultivated land, and the grazing angle is close to zero.

Estimate the maximum range at which the target can be detected, assuming that the signal-to-clutter power ratio at the receiver input must exceed −20 dB. Why, in practice, should your answer be considered very approximate?

What general effect, if any, would you expect each of the following changes to have on the range estimate?

(a) reducing the horizontal beamwidth
(b) incorporating STC
(c) increasing the pulse length
(d) increasing the transmitter power.

Q10. Derive a form of the Radar Equation for detection of targets in heavy volume clutter (equivalent to equation (3.6) in the main text for surface clutter). How does it differ from the usual form of the Radar Equation?

Use the result, together with equation (3.9), to estimate the maximum range at which a 1 m^2 target can be detected in heavy rain clutter, assuming the following parameters:

$\theta_B = 1°$ $\phi_B = 10°$ $\tau = 1\,\mu s$ $f_0 = 10\,\text{GHz}$
minimum signal-to-clutter ratio $= -10\,\text{dB}$
rainfall rate $r = 5\,\text{mm/hour}$

Why should your answer be treated with caution?

Q11. The Weibull distribution (see equation (3.11)) has often been used to describe the statistics of sea or land clutter. Show that its envelope pdf reduces to exponential form if the Weibull parameter $\alpha = 1$, and to the Rayleigh form if $\alpha = 2$.

Q12. The rms velocity spread of clutter from a forest, in strong wind conditions, is 0.5 m s^{-1}. At what frequency does the clutter power spectrum fall to one-tenth of its value at $f = 0$, if the radar wavelength is 10 cm?

Section 3.2

Q13. A terminal-area radar operates at 10 cm wavelength, with its antenna mounted 10 m above a horizontal ground surface. Find the angular separation between adjacent lobes and nulls in its vertical coverage pattern due to ground reflections.

An aircraft is flying at a constant height of 30 000 feet. How far must it travel radially to pass from one lobe centre to the next if its range from the radar is about 100 nm?

Why, in practice, is a regular pattern of lobes and nulls unlikely to be developed?

Section 3.3

Q14. The antenna of a ship's radar is mounted 20 m above the sea surface. What is its approximate distance-to-horizon for (a) a navigation buoy on the sea surface, and (b) a lighthouse tower rising to a height of 50 m above sea level? Why should the latter be much easier to detect, in spite of its greater distance?

Section 5.1

Q15. Show that the output signal from a matched filter whose input is a rectangular pulse (to which it is matched) is triangular in form with a duration twice that of the input pulse.

Q16. The pseudo-random m-sequence shown in figure 5.2(c) of the main text has successive values:

$$1, -1, 1, -1, 1, 1, 1, 1, -1, -1, -1, 1, -1, -1, 1$$

Find the autocorrelation function (ACF) of the sequence, obtained by passing it through the appropriate matched filter. What is the ratio between the central peak of the ACF and the size of its largest time-sidelobe? Why would a much longer m-sequence be used in a practical phase-coded pulse compression system?

Section 5.2

Q17. Illustrate the anti-clutter action of the log-FTC technique by a suitable set of waveform sketches.

Section 5.4.2

Q18. An MTI radar works at a frequency of 1.3 GHz with a PRF of 400 pps. The MTI system offers selectable single/double cancellation. What relative loss of signal strength is caused by MTI processing on aircraft with radial velocities of (a) 100 knots, and (b) 300 knots, using either single or double cancellation? Why might double cancellation be preferred?

Q19. Find expressions for the impulse response $h_3[n]$ and frequency response magnitude $|H_3(\Omega)|$ of a triple MTI canceller formed by cascading three single cancellers. Sketch the triple canceller as a four-pulse transversal filter, indicating the tap weights required.

Section 5.4.3

Q20. If triple PRF-stagger is incorporated in the radar of problem **Q18**, find the first blind speed for stagger ratios of (a) 5:6:7, and (b) 11:12:13. Assume the mean PRF remains at 400 pps in each case.

Section 5.4.5

Q21. A surveillance radar operates at a PRF of 500 pps. Its antenna rotates at 5 rpm and has a horizontal beamwidth of 1.8°. The double-cancellation digital MTI system uses 12-bit coding. Apart from other factors, equipment instabilities limit the MTI improvement factor I to a maximum of 60 dB.

Assuming stationary clutter, what value of I may be expected for the complete system?

Answers to Problems

The reader is asked to bear in mind that, although the following answers are formally 'correct', radar calculations generally involve a number of assumptions and simplifications.

(1 knot = 1 nm/hour = 1.152 statute miles/hour = 1.854 km/hour = 0.5150 m s^{-1})

Q1. 31.2 km.

Q2. 2332 m^2.

Q3. (a) 0.259; (b) 0.0498.

Q4. (a) 0.122; (b) 0.0174.

Q5. (a) 2.42×10^{-3}; (b) 1.49×10^{-8};
(c) 6.07×10^{-4}; (d) 1.86×10^{-8}.
$P_{fa} = 3.73 \times 10^{-6}$; about 0.36 s.

Q6. 259 nm.

Q7. +2.9 per cent.

Q8. 4.6 dB.

Q9. 24.2 km; (a) increase, (b) none, (c) decrease, (d) none.

Q10. $R_{max} = \left[\dfrac{2\sigma_t}{\eta c \tau \theta_B \phi_B \left(\dfrac{S}{C} \right)_{min}} \right]^{\frac{1}{2}}$; 4.88 km.

Q12. 21.5 Hz.

Q13. 0.143°; about 15 nm.

Q14. (a) 9.9 nm; (b) 25.5 nm.

Q16. ACF values: 1, -2, 1, 0, -1, 2, 1, 2, -3, 0, -1, -2, 1, -2, 15, -2 etc.
ratio 5:1 (14 dB).

Q18. (a) Single, -8.98 dB; double, -17.97 dB.
(b) Single, -1.04 dB; double, -2.08 dB.

Q19. $h_3[n] = \delta[n] - 3\delta[n-1] + 3\delta[n-2] - \delta[n-3]$; $|H_3(\Omega)| = 8 \sin^3 \dfrac{\omega T}{2}$

Q20. 538 knots; 1075 knots.

Q21. About 53 dB.

Index